Joining of Composite-Matrix Materials

Mel M. Schwartz

Manager, Book Acquisitions, Veronica Flint

Production Project Manager, Donna Sue Plickert

Production Coordinator, Nancy M. Sobie

The Materials Information Society

First printing, January 1995
Second printing, June 1995

This book is a collective effort involving hundreds of technical specialists. It brings together a wealth of information from world-wide sources to help scientists, engineers, and technicians solve current and long-range problems.

Comments, criticisms, and suggestions are invited, and should be forwarded to ASM International.

Library of Congress Cataloging-in-Publication Data

Schwartz, Mel M.
Joining of composite-matrix materials / Mel M. Schwartz
p. cm.
Includes index
ISBN: 0-87170-514-1
1. Composite materials—Bonding 2. Joints (Engineering)
I. Title.
TA418.9.C6S39 1994 94-32290
620.1'18—dc20 CIP

ASM International®
Materials Park, OH 44073-0002

SAN: 204-7586

Printed in the United States of America

Table of Contents

Preface

The enormous proliferation in the spectrum and sophistication of structural materials poses considerable challenges. So as not to waste the opportunities presented by new and improved engineering materials, we need new ways to incorporate materials selection, specification, processing, and use into design. Other key issues involve the ways in which quality and reliability are sought and the means by which new materials are introduced into service.

The development, selection, and use of materials have always been critical areas that underpin, and often enable, virtually all progress in engineering and technology. Users, manufacturers, and designers now have a wider range of increasingly sophisticated materials at their disposal, and many materials have the potential to be tailored for specific applications through innovative specification design and processing. Indeed, there is a revolution in materials in process.

The materials revolution can be viewed as both positive and negative. For example, the last few years have seen enormous strides in our approaches to the improvement, manufacture, selection, and use of established engineering materials. Considerable advances have been made in improving selection criteria, conducting computer-aided modeling, making selection and performance predictions, and understanding both microstructure-property relations and the behavior and demands of materials in complex engineering structures. However, there are clearly two central issues:

- The use of many new materials (e.g., brittle ceramics and anisotropic composites) requires us to move away from the established wisdom, phenomenological experience, and areas of practice that were relevant to more conventional, more homogeneous, and more tolerant engineering materials.
- In many cases, ownership of conventional raw materials will be far less strategically important than access to the best precursor materials (e.g., fibers, prepregs, whiskers, and powders) and development of the most suitable and advanced process technologies.

One measure of civilization's progress is the materials that identify successive epochs: Stone Age, Bronze Age, and Iron Age. Some materials scientists argue that the current age is the age of composites. Mainstream composites are actively changing the world, but from a manufacturing viewpoint, composites are in a tough competitive battle. Not only are material costs substantially higher, but new, more efficient production methods have had to be invented.

So why have composites endured and grown in the marketplace? Three reasons: properties, properties, properties. Even though raw materials for composite structures may cost substantially more than high-strength aluminum, composites win because they have special properties. The designers can easily specify additional carbon, glass, silicon carbide, or alumina fibers to strengthen critical areas of a structure. For example, producing a composite part using numerically controlled (NC) methods involves substantially more data than an equivalent machined part. With a machined part, you start with a big chunk of metal and then remove everything that doesn't look like the part you want. In composites, you start with nothing and proceed to construct the entire part. That's why building the part with an automatic tape-laying machine requires more NC data and the NC system has to direct the placement of hundreds of plies of reinforcement.

Much of the theoretical base of the science of composites was laid down in the 1950s, for polymer matrices, and the 1960s, for metal matrix composites (MMCs) and ceramic matrix composites (CMCs), the latter through the familiar medium of fiber-reinforced cement and concrete. After the quiescent 1970s, the 1980s saw a dramatic revival of interest in MMCs , CMCs, and intermetallic matrix composites (IMCs). The level of expenditure of research funds on these materials in Europe, the United States, and Japan has become legendary.

But the engineering problems did not vanish; they gradually intensified. The limitations on metals and ceramics for critical applications, at both low and high temperatures, inevitably mean that designers must turn again to composites. For example, even modest increases in gas-turbine engine efficiency call for significant upgrading of the materials in the engine, from front to back, and the need for fiber-reinforced metals, glasses, and ceramics and carbon-fiber-reinforced carbon becomes urgent. The resultant need for established fabrication processes (machining, drilling, cutting, joining, etc.) has thus become pressing.

Advances in joining have been deemed a highly critical need by many engineers, scientists, manufacturing joining specialists, and producers. The various chapters in this book briefly cover the variety of composite materials, their components (fiber and matrix), the fabrication and consolidation methods used, and the numerous joining procedures for similar and dissimilar composites (production, prototype, and experimental).

The use of composites in aerospace applications over the past 25 years, starting with lightly loaded secondary structures and progressing to primary wing and fuselage structures of military aircraft, have created a revolution in the design, manufacture, and use of composite materials. New aerospace systems are no longer extensions from metallic baseline designs, but rather build on the experience with composites in aircraft. Manufacturing technology, once a means to enhance productivity by automation of machining and metalworking, has matured to play a major role in the design of composite structures. The resulting emphasis on life-cycle engineering has forced designers of composite components to address the total process of materials selection, structural performance, fabrication, inspection, supportability, and repair. The challenge facing composites technology is to expand the use of composite materials to nonaerospace applications, and to develop improved materials, design concepts, processing, and manufacturing technology to achieve the full potential of advanced composite materials.

The challenge comes at an opportune time, now that we increasingly often use computer models to optimize process routes, conduct preliminary assessment of materials, analyze operational conditions, and predict performance. In the future, much preliminary screening of designs, specification of production routes, materials tailoring, and in-depth performance assessment will be carried out as computer exercises rather than by relying on prototypes and experiment. Despite this, the challenges remain of creating new standards and codes of practice for brittle, anisotropic, unforgiving materials when we still have little experience on which to build.

Composites technology is in a unique position, one in which market forces and consumer anticipation are moving at least as fast as technical innovation. The future is open to composites to be the "materials-by-design," and that they still promise to be.

Mel M. Schwartz
October 1994

Glossary of Abbreviations

ACP	advanced composite process
ADS	automated drilling system
AGT	automotive gas turbine
ATTAP	Advanced Turbine Technology Applications Project
BMI	bismaleimide
CCC	carbon-carbon composites
C-D	capacitor-discharge
CFRCMC	continuous-fiber-reinforced ceramic matrix composites
CMC	ceramic matrix composite
CTE	coefficient of thermal expansion
CVD	chemical vapor deposition
CVI	chemical vapor infiltration
DoD	Department of Defense
EBW	electron beam welding
FGC	functionally gradient ceramic
FGM	functionally gradient material
FIRE	focused infrared energy welding
GMAW	gas-metal arc welding
GTAW	gas-tungsten arc welding
HIP	hot isostatic pressing
IHPTE	integrated high-performance turbine engine
IM	injection molding
IMC	intermetallic matrix composite
LTSSB	low-temperature, solid-state bonding
MMC	metal matrix composite
NASA	National Aeronautics and Space Administration
NC	numerically controlled
PCM	polymer-coated material
PEEK	polyetheretherketone
PEI	polyether-imide
PES	polyether sulfone
PI	polyimide
PMC	polymer matrix composite
PPS	polyphenylene sulfide
PSZ	partially stabilized zirconia
RMC	resin matrix composite

RTM	resin transfer molding
SEM	scanning electron microscopy
SHS	self-propagating high-temperature synthesis
SPF/DB	superplastic forming/diffusion bonding
TLP	transient liquid-phase (bonding)
TPC	thermoplastic composite
UHMWPE	ultrahigh-molecular-weight polyethylene
XD	exothermic dispersion

Glossary of Tradenames

Astroquartz is a registered tradename of J.P. Stevens Company.

FM34 and FM34B-18 are registered tradenames of American Cyanamid Co.

Hi-Lex is a registered tradename of Hi-Shear Corp.

Huck-Comp is a registered tradename of Huck International, Inc.

Incusil ABA is a registered tradename of the Wesgo Division, GTE Products Corporation.

Kevlar and Nomex are registered tradenames of E.I. DuPont de Nemours & Company, Inc.

Noryl is a registered tradename of the GE Plastics Division, GE Company.

Plexus is a registered tradename of ITW Adhesive Systems.

Pyroceram is a registered tradename of Corning, Inc.

Torlon is a registered tradename of Amoco Chemical Company.

Vectra is a registered tradename of Hoechst Celanese Corporation.

Chapter 1

Introduction: Applications for Composite Structures

The constituents of composite materials are not naturally found together in a single material. Fiber- and particulate-reinforced inorganic materials offer many advantages over homogeneous materials in terms of weight, strength, stiffness, temperature capability, and toughness. Consequently, the development of advanced composites for structural applications has been vigorously pursued for the past 30 years. In the 1960s and 1970s, both industry and the Department of Defense/National Aeronautics and Space Administration (DoD/NASA) invested considerable resources to develop boron-reinforced aluminum using diffusion bonding technology. In the 1970s and 1980s, the main efforts were directed toward developing graphite fiber-reinforced aluminum, SiC-particulate-reinforced aluminum metal matrix composites (MMC), fiber-reinforced Pyroceram, and chemical vapor deposition (CVD) SiC ceramic composites. Although the mechanical and physical properties of these materials were exciting, they were actually used in few applications, primarily because of their high cost.

A milestone occurred when the first MMC piston insert for an automotive diesel engine was demonstrated for commercial production. Made of alumina-silica fiber-reinforced aluminum by casting, the piston insert was the first large-scale use of metal composites and was made possible by the inexpensive nature of the process and its constituents. A second milestone in the commercialization of metal composites was reached in early 1990, when a Canadian facility became capable of producing an annual capacity of 25 million pounds of its ceramic-particle-reinforced aluminum. The lesson is clear: In order to find extensive application, advanced composites must not only have much-improved mechanical properties, but also be cost-effective to fabricate. The key to this latter criterion is innovative processing methodology.

In the near future, composites will have growing application in automobiles; secondary and empennage structures of various transport aircraft; the airframe, rotors, and drive system of armed reconnaissance and light-attack helicopters; hot sections of engines and hot airframe components of military aircraft; hardware used in space vehicles and structures; and the electronics and orthopedic implant industries.

Automobiles

Even considering the cautionary predictions through the year 2000, the auto industry appears to be the hotbed for new composites applications. It has yet to tap the full potential of advanced composite materials because of their high price compared to metals, the difficulties of integrating composite part manufacturing with other operations such as painting, and the industry's generally conservative nature. But evidence of what's to come is already seen in racing cars: gearbox forks two-thirds lighter than the forged-steel parts they replace, body parts and chassis with improved energy absorption (crash worthiness), and plastic engine parts.

A sample joining problem is that plastic prop shafts must usually, although not necessarily, be coupled to an end fitting that is fabricated from metal. There are several options for producing such a joint. Two of them, the use of pins or shrinkfitted components, represent the traditional method of transmitting torque loads from one metal-based component to another. The other is bonding, which is less likely to induce stress damage in the wall of the shaft, assuming that the joint is well designed. It can be stated with some confidence that the necessary adhesives are available and that modeling, computational, and practical techniques are now sufficiently developed to allow production-line assembly of bonded composite tubes. As a result of their lightness, stiffness, and corrosion resistance these tubes are ideal for many industrial, marine, and aerospace applications in addition to the automotive power train.

Experience has shown that a joint design must take the following criteria into consideration:

- The adhesive employed must not be brittle. Preferably it should be toughened (crack resistant, not just flexible).
- For most joints, the shear modules of the adhesive required will lie in the range of 0.2 to 2.0 GPa at room temperature.
- The design of the joint must take into account the response of the adhesive to temperature changes. It should also be borne in mind that moisture absorbed during the use of the assembly will lower the modulus of the adhesive.
- If a heat-cured adhesive is used, the engaged length should be as short as possible to minimize stresses that are due to differential contraction. Normally, and fortunately, adhesives that are appropriately toughened appear to accommodate excessive stress by fracturing locally. The toughening of the adhesive prevents crack propagation and catastrophic failure.

- The joint design must not be excessively sensitive to small geometric changes, in terms of the degree to which the adhesive is stressed. This is particularly important in relation to bond line thickness and the length as a whole or the length of stepped or tapered sections.
- The bond line thickness should be as great as practically possible. This will tend to have the most significant effect on the allowable joint length.
- If possible, at least 50% of the joint should not be stressed beyond 10% of the elastic limit, to avoid creep problems. If allowed by other design considerations, 70% would be preferable.
- The adhesive should not be subjected to loads beyond its elastic limit, although there is some latitude when toughened adhesives are used. Where fatigue loads are predominantly in one direction, the ability of a joint to accept accumulated creep strain enhances the resistance of the joint to fatigue.
- Tailored joint profiles can be used advantageously. However, if the joint will be subject to wide temperature fluctuations, variation in the characteristics of the adhesive makes it essential for the fundamental geometry of the joint to be sound.

Automotive design engineers seem to agree with market analysts, 89% of whom consider metal- and fiber-reinforced plastic composites the materials most likely to see increased use in automotive manufacturing, according to one survey (Ref 1). For example, new composites technology that combines plastic and metal during injection molding (IM) is being developed for use in complex load-bearing structural components. The technology combines the cost-effectiveness and part consolidation capabilities of IM thermoplastics with the strength and stiffness capabilities of deep-drawn steel. Other benefits include weight reduction and improved fuel efficiency from replacing steel with plastic, as well as lower corrosion through selective use of steel.

Automobile Engines

The use of ceramic matrix composites (CMCs) for the hot sections of automobile engines allows the engines to operate at higher temperatures, permitting greater thermodynamic efficiency. The design of the parts are symmetrical and objects of revolution, to take advantage of the compressive strength of CMCs and minimize stress in the part. Applications have included the combustor, regenerator, turbocharger, rotor, and diffuser, and testing has been conducted at temperatures up to 1371 °C and rotor speeds up to 115,000 rpm. Processes have included slip casting and IM, while materials have included SiC, Si_3N_4, and lithium aluminosilicate (LAS) glass.

As mentioned above, the "ceramic fever" of the early 1980s has abated somewhat and has been replaced by a more cautious approach. In a survey published by Argonne National Laboratories in 1988 (Ref 2), 1990 was projected as the year for the introduction of ceramic rotors, rocker arms, cam followers, and exhaust port liners for light-duty vehicles. This survey also predicted that ceramic valve guides and valve seats would not appear until 1992 and 1993, respectively. Even further down the road (1995) would be the introduction of ceramic valves, piston caps, rings, pins, and cylinder liners. Similar predictions were made for heavy-duty vehicles, and the ceramic piston for both types of vehicles was not expected to be introduced until the year 2000. Finally, the survey predicted that ceramic gas-turbine engine applications would be introduced by 2015 and would reach a $3 to 4 billion market in 20 years (Ref 3).

Within the next three years, the following ceramic parts will be commonplace in most automobiles (Ref 4):

- Piston caps
- Cylinder and exhaust manifold liners
- Valve heads
- Turbocharger rotors

These parts will come from a variety of reinforcements, including Nippon Carbon Nicalon SiC, DuPont FP Al_2O_3, 3M Nextel 440 aluminoborosilicate, and UBE Tyranno SiC. Matrices include SiC and Si_3N_4, and densities above 90% will be the most frequently used composition.

New Engine Systems. In addition to work aimed at developing parts for conventional gasoline engines, much research is underway around the world to design entirely new engine systems that take greater advantage of the properties of ceramics and CMCs. Promising candidates include the uncooled diesel, the adiabatic or minimum-heat-loss engine, and the automotive gas-turbine engine. Turbine engines have heretofore been uncompetitive with diesel or gasoline engines in automotive uses because of the high cost of the superalloys required to withstand very high turbine inlet temperatures, insufficient material strength, and oxidation-limited engine operating temperatures. The greater temperature resistance and lower cost of ceramic materials have transformed the gas-turbine engine into a potentially viable engine for automobiles.

Government-sponsored development programs were initiated in the 1970s by several engine and automotive companies and continued through the 1980s. In addition to testing ceramic components for automotive applications, these programs developed new fabrication techniques that increased the flexural strength of the components. The use of hot isostatic pressing (HIP) found its way into the program, and new ceramic materials were also developed during this period.

References 5 to 18 describe the six-year AGT 100 project, which designed a 300 hp automotive gas-turbine engine with an all-ceramic hot section. Other literature

(Ref 2, 5-12, 19-27) describes the results of the six-year AGT 101 project, in which a 100 hp automotive gas-turbine engine featuring an all-ceramic hot section was designed for operation with a turbine-inlet temperature of 1371 °C and a rotor speed of 100,000 rpm.

Encouraged by these two programs, the Department of Energy has initiated a five-year program to continue the ceramic technology effort so that the United States will not lag behind its world competition. The purpose of the Advanced Turbine Technology Applications Project (ATTAP) is to continue to advance the technological readiness of the automotive ceramic gas-turbine engine by developing processes to fabricate and test components from next-generation ceramic materials, expand the experimental database of ceramic properties, expand the analytical tools needed to support the ceramic industry, evaluate the reliability and durability of the next generation of high-temperature ceramics, and assess low-cost ceramic processing and manufacturing technology (Ref 5, 20, 22, 23, 27, 28). The ATTAP engine, also known as the AGT-5 ceramic automotive engine, is expected to have an operating life of 3500 h at temperatures to 1371 °C. Materials and processes have been selected for some components:

- *The turbine rotor* is considered the engine's critical component because of its high-temperature requirements, geometric complexity, and susceptibility to impact damage. Initial experiments have been conducted with pressure slip casting of fracture-resistant Si_3N_4 materials for radial turbine rotors. In addition, rotor development activity has two alternative objectives: to improve IM techniques for axial turbine rotors and to toughen Si_3N_4 material systems by applying SiC-whisker reinforcement. Tests have shown that this composite material can be successfully densified by HIP with a resultant increase in strength.
- *The turbine backshroud,* a highly stressed part, will require a new type of reinforced ceramic in order to work reliably at the elevated temperatures for ATTAP testing. One process to be tried involves ram pressing of SiC that has been toughened by dispersed particles of TiB_2. Because the material is 50% stronger than direct-sintered SiC, it offers the bonus of being electrodischarge-machinable.

Thus, even if the ceramic *automobile* engine fails to materialize in the short run, researchers insist that the research and development effort will inevitably produce a far more efficient *aircraft* propulsion system. Certainly the economics are far more favorable (Ref 29). Aircraft engines are typically produced in low volume at high prices, allowing the elevated cost of ceramic components to be easily absorbed. According to Patiky (Ref 30), an automobile engine costs just \$3/hp, while a jet engine costs \$250/hp because of the far finer engineering tolerances required. As a result, material costs represent a much lower fraction of aeronautical costs. In addition, the thrust increases possible in a ceramic jet engine mean a lot more power for the money. Engineers thus forecast that ceramic engines will be powering planes within 15 years.

Materials and Processing Considerations. Fiber-reinforced CMCs have improved resistance and hence more potential for defect tolerance. Consequently, they have received a great deal of attention for use in high-temperature structural applications. A reinforcing phase having the proper mechanical, physical, and chemical properties can introduce internal deformation and fracture mechanisms that decrease the sensitivity of the matrix to flaws and increase the fracture toughness of the ceramic. As with other composite systems, the reinforcing ceramic second phase can have a variety of shapes, ranging from nearly spherical particles through whiskers and chopped fibers to effectively continuous fibers. Compact particles and whiskers lend themselves to conventional ceramic processing methods. In addition to toughening parts in all directions, this reinforcement approach is relatively inexpensive, and thus it is well suited to mass production (Ref 31-38) (Fig. 1-1).

Many companies have continued to develop commercially available high-performance fibers, with emphasis on Al_2O_3, SiC, Nextel, Nicalon, and others (i.e., SiC fibers produced by CVD). Such fibers have significantly greater tensile strength

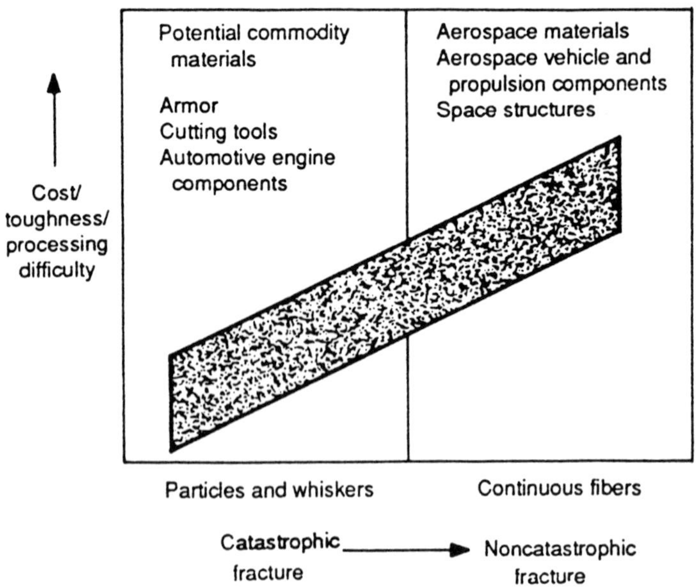

Fig. 1-1 The two categories of ceramic matrix composites, based on failure mode and processability. Source: Ref 31, 34

and stiffness than continuous strands produced by other methods, such as conversion of ceramic precursor polymers to fibers followed by pyrolysis to accomplish conversion to a ceramic (Ref 39, 40). Other processing approaches include those in which fibers are made into shapes, matrix precursor is added, and the combination is heated to a high enough temperature to turn the precursor into a ceramic matrix, but low enough to preserve the fibers. Three-dimensional fiber shapes can be formed by weaving, two-dimensional shapes can be formed by tape lay-up, and bodies of revolution can be formed by either method.

Matrices have been made by reaction bonding, polymer pyrolysis, CVD, or sol-gel processing (Ref 41-46). In reaction bonding, alternate plies of fiber tape and silicon sheet are laid up, then heated in nitrogen to convert the silicon to Si_3N_4 and bond the layers together (Ref 47, 48). Chemical vapor deposition slowly builds up the matrix on the surfaces of the fibers until the spaces between them fill up. The remaining two methods, polymer pyrolysis and sol-gel processing, essentially consist of adding liquid and heat treating (Ref 31, 39, 49-62).

Reaction-based processing offers promise even though it is not as advanced as conventional CMC processes. It appears uniquely capable of fabricating parts for high-temperature applications because of the extremely high purity levels and unusual phase chemistries that can be achieved. For low-temperature applications, reaction-based processing is particularly attractive for making large, complex, net-shape, and net-dimension parts that are difficult or expensive to make with conventional densification-based ceramic processes. Recently developed process models are expected to improve the ability to make complex shapes.

Aircraft

Air Force efforts on the F-111 empennage started the advanced composites age in aircraft. It is an interesting side note that repair of cracked metal structure also started back then with boron composites.

The use of progressively more composite on each new aircraft occurred after this time. The F-14 and F-15 both had boron-epoxy tails. The F-15 also had carbon-epoxy in a speed brake. In the next set of military fighter aircraft, the F-16 and F-18, the composite quantities grew. The entire tail of the F-16 was carbon-epoxy, and the F-18 had additional composite structures, including wing skins. Next, the AV-8B Harrier had its entire tail and wing made from carbon-epoxy, in addition to significant fuselage structure. The composites weight in the AV-8B was about 26% of the structure weight. Subsequent fighter aircraft will get very close to the prediction of the maximum weight possible on fighters. The composites age in fighter aircraft is here.

In general aviation, the composites age has arrived at some firms. The Beech Starship is one that is using all-composites (Ref 63-80).

Transport aircraft, which includes large military aircraft and commercial airliners, are not as far along with the use of composites, but again, composites are being used in ever-increasing percentages. The tail structure of the AirBus and the projected tail of the Boeing 777 will be carbon-epoxy, and other composite structures are being developed for the large aircraft. NASA has sponsored the development, certification, and flight service evaluation of composite secondary and empennage structures (Table 1-1).

Spoilers for the Boeing 737, one of the components in the NASA flight service program, have been in service since 1973, and tests after 15 years of service indicate no loss in residual strength (Ref 81). This experience made possible the production commitment of secondary composite structures on the Boeing 767 and the baseline designs of numerous components for the McDonnell-Douglas MD-11 and C-17 (Fig. 1-2). However, full exploitation of the benefits of composites requires application to wing and fuselage structures, which account for approximately 75% of the aircraft structural weight.

Obstacles to increasing the use of composites in aircraft structures have included the lack of an integrated database with a maturity similar to that for metals, the high initial cost of materials, and the lack of automated affordable manufacturing processes. The major obstacle, however, has been acquisition cost. In the mid-1980s, data indicated that the cost of composite structures was substantially more than the cost of similar metallic components (Fig. 1-3). Yet composite structures designed as replacements for metallic structures contain significantly fewer parts, and numerous studies have noted the direct relation between part count and cost. Projections for the acquisition costs of composite structures in future production range from 10 to 40% less than those for metallic structures (Ref 82).

For 21st-century aircraft, polymer matrices and carbon-fiber-reinforced composites must exhibit a favorable balance among such properties as toughness,

Table 1-1 Composite structures developed under NASA AEEE program

	Secondary structures			Empennage structures		
Component	DC-10 rudder	727 elevator	L1011 aileron	DC-10 vertical stabilizer	723 horizontal stabilizer	L-1011 vertical fin
Size (root × span), ft	3.2 × 13.2	3.4 × 17.4	4.3 × 7.7	6.8 × 22.8	4.3 × 16.7	8.9 × 25
Metal design weight, lb	91	130	140	1005	262	858
Composite design weight, lb	67	98	107	834	204	622
Weight reduction, %	26	25	24	17	22	28
Number of production units	20	11	12	3	11	2
Month and year of first flight service	4/76	3/80	3/82	1/87	3/84	...

Source: Ref 1

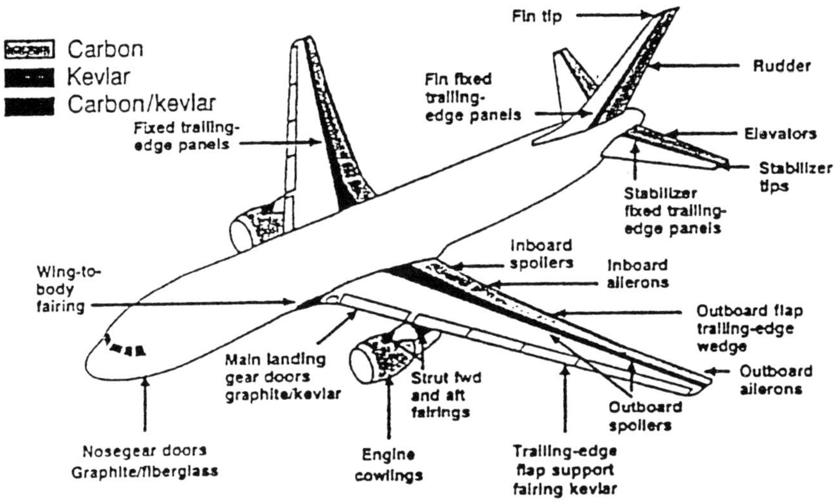

Fig. 1-2 Current uses of composite in Boeing commercial airplanes. Source: Ref 108

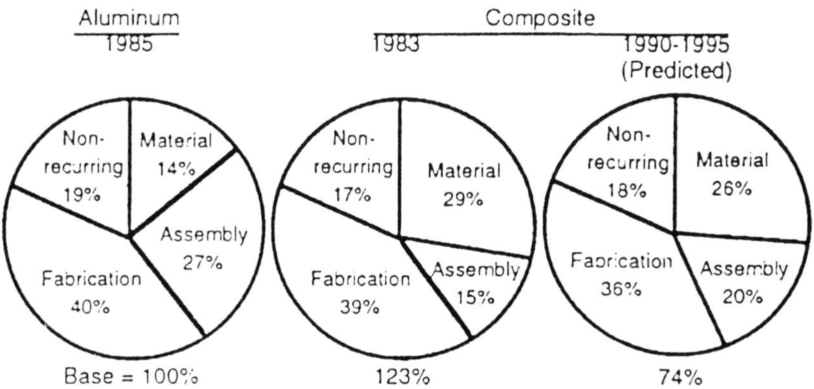

Fig. 1-3 Comparison of costs for aluminum and composite wing structures in commercial aircraft. Source: Ref 109

damage tolerance, processability, good mechanical properties under hot/wet exposure, and acceptable cost. The major resin matrix for almost all the current secondary and medium primary subsonic aircraft structure in service today is based on a brittle 177 °C curing epoxy system. These brittle matrices in conventional composite form do not exhibit the necessary toughness and damage tolerance required of heavily loaded primary aircraft structures. Consequently, new 177 °C cure-toughened epoxy composites have been developed to overcome the damage tolerance problem. The hot/wet unidirectional (0°) compressive strengths of these materials, and the compressive strengths after impact, equal or exceed typical material specifications and far exceed those of first-generation brittle epoxy composites (Ref 83, 84). However, prepregs of these toughened materials are at least three times more expensive than standard brittle systems and are considered too expensive for widespread use on commercial aircraft (Ref 85).

Thermoplastic composites have received considerable attention over the past six years because they offer an attractive combination of mechanical properties, toughness, and the potential of low-cost manufacturing (Ref 86, 87). Relative to thermosets, they offer simple formulations, excellent prepreg stability with infinite out-time and no cold storage requirements, and fast thermoformability. A few of their more obvious drawbacks include tackless/boardy prepreg, high processing temperatures, poor fiber alignment, fiber waviness, relatively poor 0° compressive strengths, difficult prepregging, and a small database.

As with toughened thermosets, thermoplastic prepreg is much more expensive than the standard brittle epoxy prepreg. Thus, efforts are underway to develop affordable thermoset and thermoplastic composites for subsonic aircraft. They involve textile preform technology, resin transfer molding (RTM) materials and fabrication technology, powder-coated towpreg, tape prepreg fabrication, and commingled towpreg fabrication. These efforts show exciting results:

- Net-shape fiber preforms can be infiltrated by RTM processes and cured by conventional means (Ref 88-90). Automated textile processes such as through-the-thickness weaving, braiding, knitting, and stitching offer the potential for reduced fabrication costs of aircraft structural components. Complex fiber architectures, multilayer fabrics, and structural composite preforms have already been demonstrated using automated textile technologies (Fig. 1-4). The preforms lend themselves to economical composite fabrication techniques such as RTM and pultrusion.
- The technique for constructing stitched dry preforms from uniweave fabric subelements is briefly illustrated for a reinforced panel in Fig. 1-5. The resin can be either vacuum or pressure infiltrated into the preform, which is bagged and supported by a die consisting of easily fabricated plate metal inserts. The lay-up is cured using standard autoclave procedures.
- Designers are recognizing that the cost of a structure is influenced by fabrication and assembly costs, not just material cost. Design concepts that reduce fabrication and assembly costs can often be cost-effective with more expensive materials.

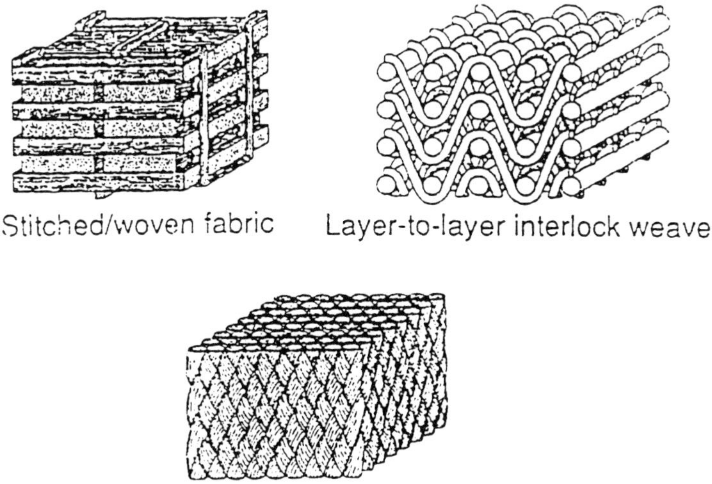

Fig. 1-4 Through-the-thickness reinforced textile forms. Source: Ref 89

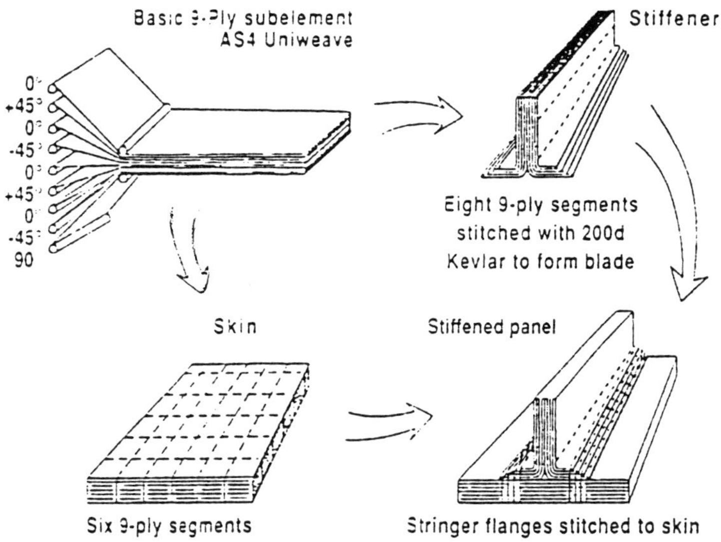

Fig. 1-5 Technique for constructing damage-tolerant stiffened panel from uniweave fabric. See text for detail. Source: Ref 110, 111

Lower-cost structures usually involve simpler design concepts that reduce the number of parts or elements that have to be assembled. For example, several structural elements can be integrated into a structure during subcomponent fabrication by co-curing the stiffeners with the skin. This significantly reduces the number of fasteners, clips, and shear ties needed to assemble conventional metallic stiffened structures. "Sandwich" structures are being considered because they have the potential to minimize or even eliminate the need for stiffeners.

Design concepts are being developed that exploit the advantages of new and existing fabrication procedures to reduce costs. For example, automated advanced fiber placement machines can be used to fabricate integrated structures by rapidly winding or placing fibers or tows (untwisted bundles of continuous fibers) onto fabrication tools or mandrels. Stiffeners can be wound into grooves in the mandrel that are oriented in any suitable stiffener orientation, such as a geodesic stiffener pattern, and then the skin can be wound over the stiffeners to produce a stiffened shell structure with minimum part count (Fig. 1-6). Resin transfer molding with textile preforms is another example of a fabrication process that can be used to fabricate integrated structural components.

The current challenge is to develop and verify manufacturing processes that will yield structures that are both cost-effective and reliable. Numerous lessons have been learned in the past 20 years about the costs of fabricating and nondestructively inspecting composite structures for civil aircraft. Reference 91 is an excellent discussion of cost-estimating processes for composite structures. Developments believed to be critical to achieving cost-effective structures include:

- Lower-cost starting materials that do not add to the downstream manufacturing cost

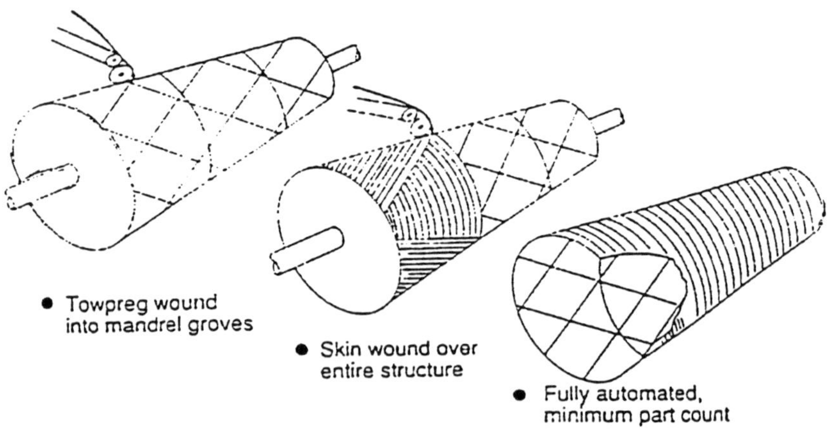

Fig. 1-6 Fabrication of an automated fuselage. See text for details.

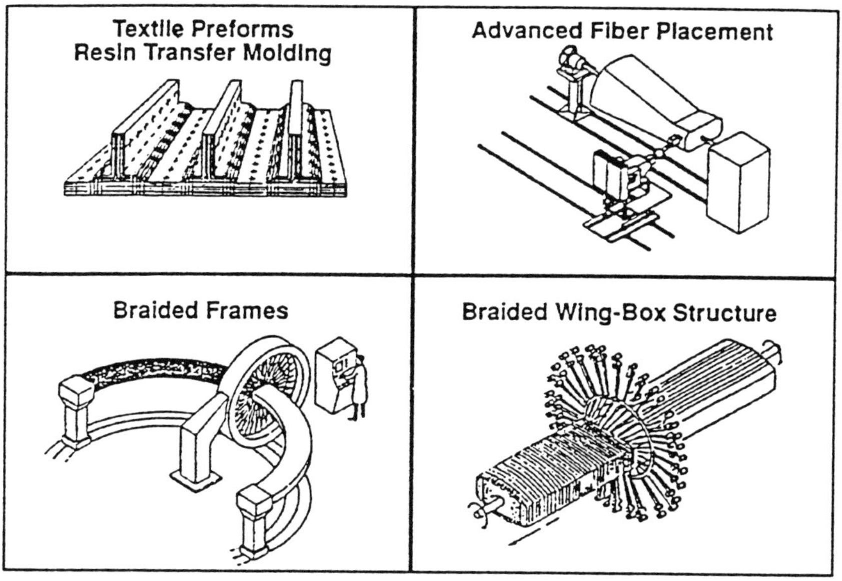

Fig. 1-7 Processes for assembling composite structures for commercial aircraft.

- Significantly more automation, to reduce labor cost and eliminate scrap due to human error
- Simple designs that exploit automated processes
- Elimination of the autoclave consolidation
- Assembly processes that require shimming

Specific assembly processes currently being developed include RTM, advanced fiber placement, and braiding (Fig. 1-7). Textile preform methods are another relevant development, including through-the-thickness stitching of uniweave tape and two- and three-dimensional weaving and braiding of structural shapes (Ref 92-95).

Helicopters

Helicopters were probably the first aircraft to incorporate composites in primary structures. Helicopters with main rotor blades made of unidirectional glass-epoxy were flying in the late 1960s. (The design of a rotor blade is based on stiffness; weight savings is not a criterion.) From this start, helicopter manufacturers have continually designed more composites into their vehicles. The Model 360 and the V-22 Osprey have put the vertical lift aircraft fully into the composites age.

For the newly-designed U.S. Army RAH-66 Comanche helicopter, increased use of composite materials was necessary to meet aggressive performance and increased militarization requirements. A system-design approach, coupled with an integrated product development philosophy, was used to find solutions predicated on the use of composite materials.

Major design innovations, coupled with new materials and manufacturing technology, increased the use of composite material systems to a level exceeding 50% of all materials used (Fig. 1-8). In the airframe subsystem, composites are used in the skins, doors, stringers, frames, bulkheads, internal center keelbeam box structure, main rotor pylon fairing, Fantail shroud, vertical pylon, and horizontal stabilizer. In the rotor subsystem, composites were used in the flexbeam, blades, torque tube, rotating swashplate, quill shaft, and main rotor fairing. In the drive subsystem, composites were used in the drive shafts and main transmission housing. The total use of composite materials in the airframe, rotors, and drive subsystems is unparalleled in previous helicopters.

The major composite materials used included new toughened epoxy resins, bismaleimide resins, graphite fibers with improved stiffness and strain, and fiberglass and Kevlar fibers (Fig. 1-9). The greatest growth was in the use of graphite-epoxy, and as a result, new manufacturing and design solutions evolved, such as the rotor flexbeam and airframe keelbeam center box and the split-torque main

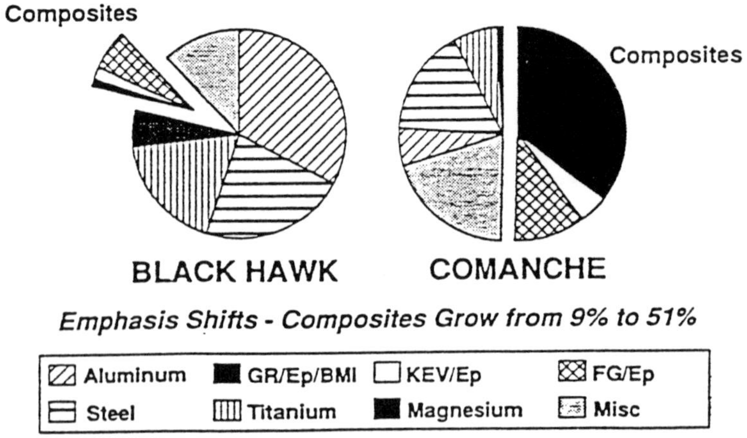

Fig. 1-8 Materials used in the U.S. Army Comanche helicopter compared to materials used in a previous design. Gr/Ep/BMI, graphite-epoxy-bismaleimide; KEV/Ep, Kevlar-epoxy; FG/Ep, fiberglass-epoxy

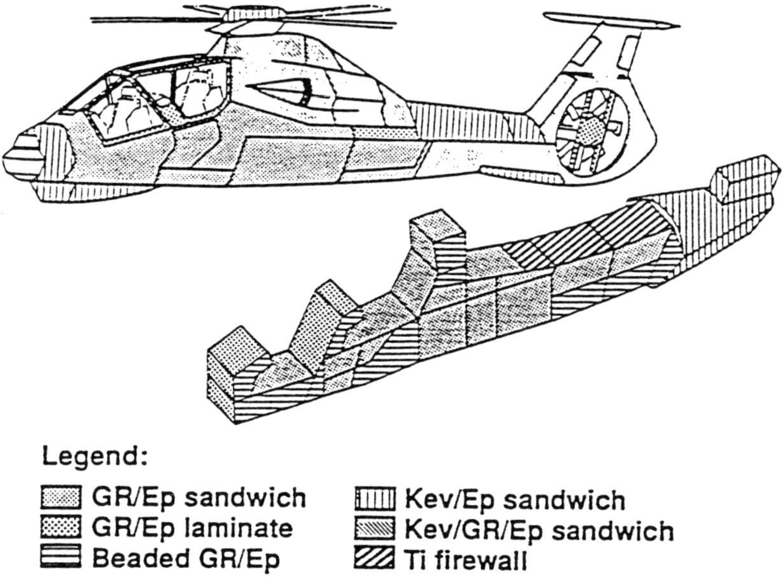

Fig. 1-9 Materials used in the airframe design of the U.S. Army Comanche helicopter. GR/Ep, graphite-epoxy; Kev/Ep, Kevlar-epoxy; Kev/GR/Ep, Kevlar-graphite-epoxy

transmission gear box housing composite subassembly. The rotor design (Fig.1-10) incorporated parts made using graphite-epoxy and glass-epoxy prepreg broadgoods materials and made use of filament winding, braiding, and RTM. The main elements of the drive system include an advanced-technology split torque main gearbox, a ballistically tolerant tail drive shaft, and a single tail gearbox. Advanced composite materials used included graphite-epoxy filament-wound tail drive shafts, graphite fiber preforms, and bismaleimide resin systems used in RTM for main gear box housings.

Engines and Hot Components—Military Aircraft

Hypersonic aircraft have created the need for the integrated high-performance turbine engine (IHPTE). Present engines operate at temperatures up to 982 °C, whereas the IHPTE will require temperatures of 1649 to 1927 °C. At the same time, it will have to endure high pressures and tolerate impact without catastrophic failure. Large amounts of CMC, MMC, and CCC (carbon-carbon composites) will be contained in hypersonic vehicle propulsion systems; by the year 2010, as much as 80% of these engines could be CMC and MMC (Fig. 1-11). Examples include

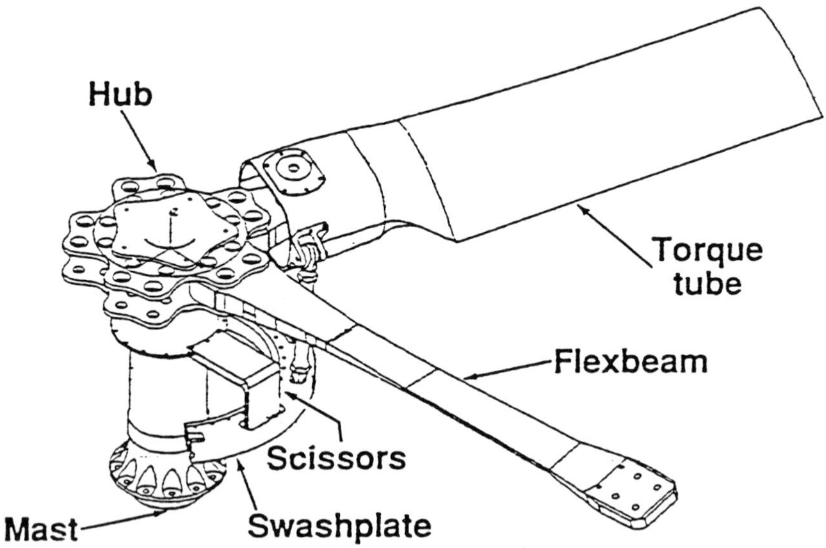

Fig. 1-10 Rotor design of the U.S. Army Comanche helicopter.

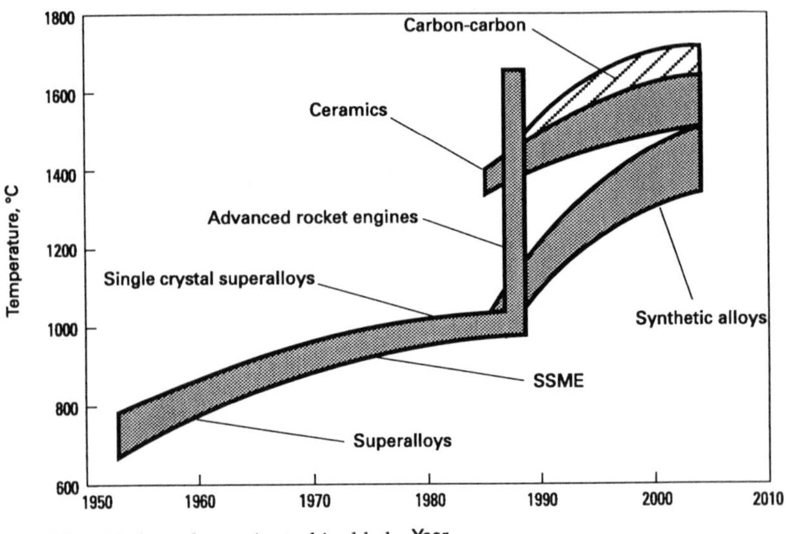

Fig. 1-11 Materials for rocket engine turbine blades.

SiC-reinforced carbon compressor blades and CMC ramjet combustion chamber rocket exhaust nozzles (Ref 96), along with controls for aerospace planes (Fig. 1-12). Another example is that the flight of a Mirage 2000 fighter aircraft has validated the use of a CMC from the French company SEP (Ref 97). With the use of chemical vapor infiltration (CVI), SEP made up half of the "hot" and "cold" flaps for the engine, whereupon SiC fibers in a SiC matrix were incorporated into the engine flaps. Finally, look for thrust-vectoring nozzles, hollow blades, and magnetic bearings in tomorrow's gas turbines to be made from ceramic composite materials.

Advantages. The main advantage of C-SiC composite nozzles for large liquid-fueled engines is their capability to retain their mechanical characteristics up to 1800 °C, eliminating the need for a nozzle cooling system. As a result, the nozzle is lighter than metallic units that require cooling. Other expected advantages of C-SiC nozzles include their low erosion rates, high resistance to oxidation, very high shear strength, and resistance to thermal shock.

In a study recently completed by several engine manufacturers and government agencies, it was found that continuous-fiber-reinforced CMCs (CFRCMCs) appear to have great potential for use in rocket engine turbine components. The

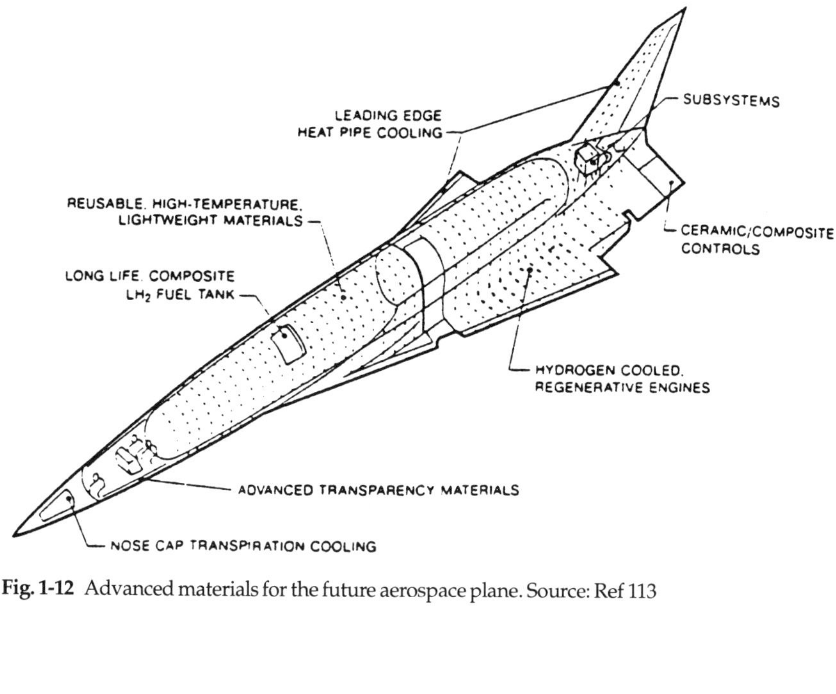

Fig. 1-12 Advanced materials for the future aerospace plane. Source: Ref 113

fiber-matrix systems tested were SiC/lithium-alumino-silicate, SiC/magnesium-alumino-silicate, SiC/calcium-alumino-silicate, SiC/black glass, SiC/borosilicate, SiC/silica, C/lithium-alumino-silicate, C/borosilicate, C/silica, C/alumina, SiC/silicon nitride, SiC-SiC, C-SiC, and C-C. The screening criteria were maximum operating temperature, thermal shock resistance, environmental resistance, ultimate tensile strength, fracture toughness, fabricability, and maturity. The study confirmed that the use of CFRCMC materials in the severe environment of the rocket engine turbopump hot section offers three distinct benefits: increased performance, improved component life, and greater design flexibility (Ref 97). C-SiC was the most promising material because it exhibited the necessary combination of mechanical properties, ease of fabrication, and adequate environmental resistance.

In a series of tests on a small H_2-O_2 rocket engine, only SiC-SiC and SiC-Si_3N_4 demonstrated any ability to survive the severe thermal shock environment. These tests produced some evidence that CFRCMCs exhibit many of the properties required for rocket engine turbines, including low density, environmental durability, thermal-shock resistance, toughness, and reliability. There is great potential for CFRCMCs to increase performance and payload capabilities, improve component life, and improve design flexibility.

The national aerospace plane, or X-30, will take off and land on conventional runways, cruise at approximately 27,000 km/h, and reach orbit in a single propulsion stage. The most critical factor in achieving these goals is the development of suitable lightweight, high-strength materials that will perform satisfactorily at much higher temperatures than any of today's materials and structural components (Ref 98).

Because SiC-fiber-reinforced titanium-matrix composites retain strength and stiffness up to 980 °C, titanium-matrix composites are being investigated for the national aerospace plane. The research includes matrix development and characterization, fiber development, processing, structural fabrication, and component testing. Processing methods for composites include fiber-on-foil, plasma spray, and low-pressure plasma spraying techniques. The most desirable matrix properties are ambient-temperature ductility and high-temperature interfacial compatibility in the fiber, coupled with higher strength, low density, and ease of fabrication.

In studies of laser beam welding and diffusion bonding of SiC and Ti-6Al-4V to each other, an adequate joint was obtained in both processes by using Ti-6Al-4V as filler metal (Ref 99-101). These processes can be used to join MMCs to monolithic metals, which is important because combined use of MMCs and monolithic materials will be expected in practical components. Laser beam welding of SiC/Ti-6Al-4V to Ti-6Al-4V plate has also been performed (Ref 102). By controlling the laser beam so that it falls on the optimum position on the Ti-6Al-4V plate, a joint was achieved that had a strength comparable to the tensile strength of the Ti-6Al-4V base metal.

The following results were obtained in diffusion bonded, continuous-SiC-fiber-reinforced Ti-6Al-4V composite:

- Bonding strength increased with bonding time, to a maximum of 850 MPa for a seven-ply composite with fiber volume of 30% [V_f (30)] and 650 MPa for a ten-ply composite with fiber volume of 45% [V_f (45)]. In V_f (30) composite the maximum joint strength was obtained for a bonding pressure of 7 MPa and a bonding time of 7.2 ks. The maximum joint strength was 85% of the Ti-6Al-4V base metal tensile strength.
- The void ratio at the interface between the composite matrix and Ti-6Al-4V decreased with increasing bonding time. The bond strength between fiber and Ti-6Al-4V plate may not contribute to the joint strength in this condition. The joint strength was controlled by the bonding of the composite matrix and the Ti-6Al-4V plate.
- The bondability of SiC/Ti-6Al-4V composite to Ti-6Al-4V alloy was better for V_f (30) composite than for V_f (45) composite. The reason is that the bonding pressure for each fiber is higher, and the distance from one fiber to another is greater, in V_f(30) composite than in V_f (45) composite, so plastic deformation of Ti-6Al-4V alloy occurs more easily in V_f (30) composite.
- In high-strength joints the fiber side was debonded from the composite matrix. At the bond interface, the state of stress may be triaxial because of the restraint by fibers and/or the defects produced around the fibers. This may explain why the joint strength was 100 to 150 MPa higher than the value simply calculated from the area fraction of the bonded matrix/Ti-6Al-4V interface.

The prime candidates for materials with high specific creep strength are titanium aluminides and SiC-filament-reinforced titanium-matrix composite. The important fibers are titanium diboride and alumina, which have coefficients of thermal expansion (CTEs) closer to that of titanium than do SiC fibers.

The national aerospace plane will have numerous components (e.g., leading edges and combustion chamber walls) that will require extensive cooling of the surfaces. Several materials are under development for high thermal conductivity and structural efficiency at elevated temperatures. One is a copper-matrix composite reinforced with high-modulus graphite fiber for good structural and thermal properties. The research efforts are directed to infiltrate graphite tows to coat individual filaments with a layer of copper and then compact the composite. Another copper composite under development has a discontinuous second phase, such as niobium, for better elevated-temperature directional strengths. Beryllium combines low density, high stiffness, and good thermal conductivity. Research is continuing on methods to increase structural efficiency at temperatures up to 760 °C.

Electronics

Conventional ceramics have enhanced many important aspects of modern technological society. The next generation of ceramics, advanced ceramics and ceramic composites, promises to enable the development of still-higher-performance devices. These ceramics and ceramic composites can be tailored to have unique combinations of electrical, optical, mechanical, and chemical properties that make them essential and irreplaceable in many engineering applications.

At present, electronics applications dominate the world market for high-technology ceramics. Ceramics for electronics applications are high-band-lap, largely insulating ceramics and glasses in which a broad range of dielectric, elastic, optical, thermal, electrical-conductive, piezoelectric, and pyroelectric-tensor properties can be manipulated and controlled to close tolerances. In most applications the ceramic does not stand alone, but rather is bonded to a metal (for electrodes and conductors) or polymer, which requires precise control and understanding of the interface. Often the metal and ceramic parts are in closely spaced lamellar structures, as for multilayer interconnection packages or multilayer capacitors, and must be coprocessed through the whole ceramics forming and firing process to yield defect-free monolithic hermetic packages.

For integrated-circuit packaging, glass-bonded Al_2O_3 ceramics are widely used. The green ceramic is usually tape cast with a suitable organic vehicle, and molybdenum metallization inks are screen printed for the interconnection wiring. Automation is used extensively in inspection, punching, printing, and lamination. However, although Al_2O_3 is the dominant, traditional ceramic substrate material, its low thermal conductivity makes it vulnerable to replacement with other materials that can dissipate heat from an electronics package more rapidly. Despite their much higher cost, SiC, AlN, and BN are challenging Al_2O_3, particularly for the more critical applications (Ref 103). BeO, the second most common packaging material, is also threatened by some of the newer materials (Ref 104). AlN in solid shapes, and AlN and SiC in substrates, are nontoxic, pressureless sintered ceramics with excellent heat dissipation and electrical resistivity characteristics. They have a mechanical strength 1.5 times greater than that of Al_2O_3 or BeO. AlN is taking the place of BeO in some substrates and multilayer semiconductor packages, and materials such as AlN and BN are replacing BeO in power-transistor mounting wafers and conductive epoxy adhesives.

In the past, tape casting has been used to produce thin layers of ceramic-loaded polymers that can be stacked and laminated into multilayered structures. Today, tape casting provides the products that are the backbone of two $1 billion industries: multilayered capacitors and multilayered ceramic packages. The basis for these products is the ability to cofire or cosinter the cast tape and the metallization simultaneously. This new process technology, multilayer polyimide (PI), permits the design of high-density structures using microwave-like construction, or burying signals between ground layers with typical lines less than 25 μm wide and with 60 μm pitch. By alternating several layers of PI films onto a ceramic, this process can be used to produce state-of-the-art multichip modules. It can also be used as a fundamental technology in high-speed integrated circuit interconnection and single-die packaging (Fig. 1-13).

A simple modified tape-casting procedure has been developed for application to ceramic joining when the joining materials are in powder form (Ref 105). It involves preparing a slurry from the powder, a solvent, and a thermoplastic binder, then casting directly onto the joining surface using a moving doctor blade. Handling of the tape prior to joining is not necessary. Thus, the binder content is minimized, plasticizers are not required, and the viscosity is controlled by solvent content. The utility of this method for producing joints with thin uniform interlayers has been demonstrated using hot-pressed SiC joined to itself with TiC + Ni powder material. The joint thickness was approximately 75 mm.

Polymer packaging engineers have longed to apply the low dielectric constant of easily processed polymers to high-speed circuit substrates. Anisotropic ceramic-fiber laminates now offer that opportunity. Electronics industry trends toward denser and faster circuits are placing demands on microelectronics packaging, encouraging the development of new dielectric materials for improved heat removal, lower dielectric constant, and improved reliability. Recent studies (Ref 106, 107) indicate that a composite of ceramic fibers and polymers can achieve an excellent property balance for electronics. It offers the high thermal conductivity

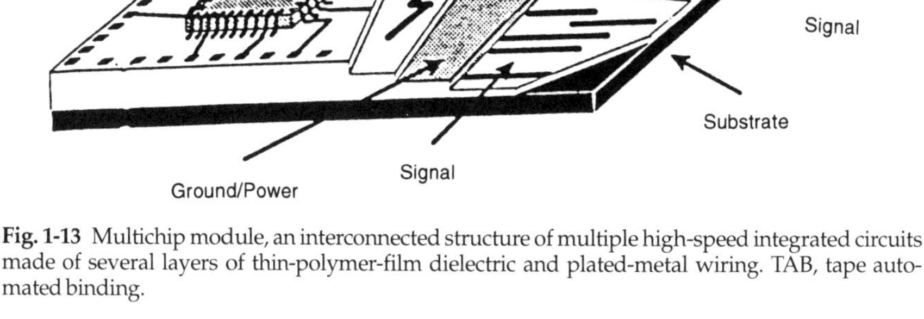

Fig. 1-13 Multichip module, an interconnected structure of multiple high-speed integrated circuits made of several layers of thin-polymer-film dielectric and plated-metal wiring. TAB, tape automated binding.

and stability of ceramics coupled with the low dielectric constant and ease of processing of polymers. By combining these materials, the orientation and morphology of the ceramic filler can be manipulated to provide a favorable anisotropy, a directionality for key properties, in the microelectronic package. This is a primary consideration in the composites approach that uses ceramic-fiber composites with Al_2O_3 spherical particles in PI thin films.

Orthopedic Implants

The role of fiber-reinforced plastics in aerospace and automotive applications has been closely followed by orthopedic implant manufacturers. High-performance metal alloys are used in orthopedic implants, and the industry might be on the road to repeating the same with advanced composite materials. The recent emergence of high-performance polymers and high-strength fibers has made available a new class of engineering materials for a multitude of product applications.

Advantages. Composite materials have several advantages over metals when used in orthopedic implants:

- High-performance fibers and engineering polymers have become more cost-competitive in recent years, while specialty-metal alloys have not shown a similar trend. The metals that are widely used in orthopedic implants are cobalt-chromium and titanium alloys. The cost of these alloys and the associated machining costs have only shown an upward trend with time, while with polymer-fiber composites one could take advantage of efficient, fast continuous processes such as pultrusion and IM.
- Composite materials give the design engineer greater flexibility in optimizing final product performance. By orienting the reinforcing fibers in various optimum directions, one can control engineering properties such as strength and modulus, according to the performance requirements of the implant.
- Because of the lower modulus of the composite material, the stress transfer at the implant-bone interface is more uniform and efficient. This might lead to better bone remodeling and longer implant service life.
- To date, thermoplastic polymer/fiber composite materials have exhibited excellent tissue biocompatibility in animal studies. Polymer-fiber composites are expected to end metal-ion-induced reactions in some sensitive patients.

Challenges. Despite the advantages, composite materials have not been used extensively in orthopedic implants, because:

- The cost savings has not been obvious.
- The tremendous investment in metals-based technology has stifled any significant attempt to switch to composite materials.
- There have been limited uses for composite materials in highly specific applications.
- Few technical personnel are skilled in designing and working with composites.

Design. Composite materials offer tremendous design freedom. The design engineer can choose from a broad array of polymer and reinforcing fibers to create implants that outperform their metallic counterparts in cost and function. For example, a typical hip prosthesis is made of a high-strength, high-modulus metal alloy such as cast or forged Co-Cr and Ti-6Al-4V alloys. The implant is fixed into the femoral canal using an acrylic bone cement. The major causes of failure of hip prostheses are implant loosening and fracture, and composite materials offer a new dimension in addressing these problems.

Product design must be thoroughly considered at an early stage. To finalize a valid design, a highly interactive team is needed whose members have expertise in implant design, composite material design, polymer and fiber selection, testing methodologies specific to advanced composite structures, material biocompatibility, and composite manufacturing methods. The design must have the final characteristics:

Biocompatibility. The final material must be extremely biocompatible.

Implant Stability. The composite material must be stable at the service environment for the full length of the expected service life. For implantable devices, hydrolytic stability is one of the primary concerns. Thermoplastics exhibit better hydrolytic stability than thermosets.

Material and Processing Costs. The composite materials and their associated processing techniques must be cost-effective. In this respect, thermoplastic polymer/fiber composites appear to have the edge. Continuous processes such as pultrusion, pulforming, or extrusion along with fast batch processes such as IM, thermoforming, and stamp molding produce high-quality, inexpensive end products.

Matrix and Fiber Relationship. Thermoplastics may serve as the matrix of choice for high-performance orthopedic implants. The geometry of the reinforcing fiber also plays a dominant role in the final mechanical properties of the composite materials. Obviously, a randomly oriented reinforcing fiber will not impart the same mechanical properties as a continuous, unidirectional fiber. In the former case, the material behavior is almost homogeneous and isotropic, while the latter is highly anisotropic. One should resort to advanced composites only if the mechanical properties for the final product are reasonably anisotropic.

Computer-Aided Design. Orthopedic implants such as hip joints and bone plates have to be designed for high specific strength in localized geometries. The primary challenge for the composite materials designer is to achieve a final fiber-matrix arrangement in which the fibers are the principal carriers of the applied stress. Because of the highly anisotropic behavior of advanced composite materials, the principal stresses in an implant have to be carefully identified and the fibers oriented accordingly. Finite element modeling techniques could be extensively used to predict the implant performance in vivo.

Testing methods used to evaluate metals and metallic implants may not be directly applicable to the testing of composite materials. Two identical samples of composite materials tested in slightly different orientations could produce dramatically different types of failure. Recently, three new state-of-the-art materials have been developed:

- *Polished Al_2O_3 ceramic*
- *Carbon-fiber-reinforced ultrahigh-molecular-weight polyethylene (UHMWPE):* Random fibers were molded into UHMWPE to improve the creep resistance and the strength. Because carbon is a well-accepted material from a biocompatibility standpoint, and because UHMWPE has proven itself for many years, this combination is expected to perform well as an effective, biocompatible articulating surface. The material is still in its infancy as far as being used in orthopedic implants.
- *Carbon-fiber-reinforced phenolic thermoset resin:* A material that has only very recently been considered as an articulating surface against polished ceramic is a phenolic thermoset resin filled with semioriented short carbon fiber, called TCF. TCF possesses creep and strength properties that are superior to those of UHMWPE. Wear testing against polished ceramic has shown that the wear rate is similar to that of ceramic on ceramic and that the frictional moment actually decreases with time. Successful biocompatibility testing has been completed, and this composite material may be available in the United States within the next few years.

The frictional characteristics of the three materials are shown in Fig. 1-14.

Composite implants will most likely be developed initially for reconstructive devices, especially in the hip. Because composites using high-strength fiber impregnated with polymer resins can most easily accommodate the combined modulus and strength requirements, it is expected that this type will be pursued first. Metal matrix composites do not represent the same design flexibility. CCCs must be relatively porous to exhibit a low modulus, and this porosity may result in a less than optimum fatigue strength.

It will undoubtedly take a significant research effort to answer all the concerns about the use of composites in orthopedic implants. However, because the bone itself is a composite structure, well-designed composite implants may ultimately best accomplish the intended function. Certainly the state-of-the-art structural implant materials of the future will include composites.

Space

Because of the large temperature variations in space (–160 to +93 °C) and the need to maintain precise alignment of communication and sensor systems, dimensional stability is a critical factor. Composites reinforced with graphite and Kevlar aramid fibers possess a high specific strength and modulus and a low CTE, which makes them very attractive for spacecraft applications. Stiffness and low thermal distortion, rather than strength, are the major considerations for spacecraft, so the material properties needed will differ from those used in aircraft. Other important factors for spacecraft include thermal and electrical conductivity, long-term stability under vacuum, space radiation, and low outgassing.

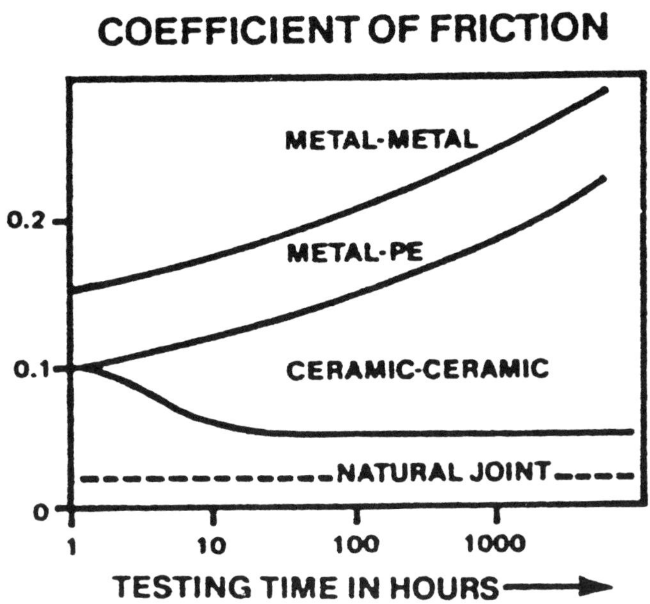

Fig. 1-14 Frictional characteristics of orthopedic implants and human joints. PE, polyethylene.

Dimensional stability alone greatly favors graphite-epoxy and Kevlar-epoxy composites because of their high stiffness and very low CTE. Kevlar and graphite fibers exhibit negative CTE and epoxy exhibits normal CTE, so the resultant composite has no thermal expansion or contraction, even over a large temperature range. For example, consider the graphite-epoxy "mechanical arm" of the space shuttle, which is a tubular boom 38 cm in diameter and 15 m long. It contains wrist, elbow, and shoulder joints for rotating in the yaw, pitch, and roll modes. On earth, this mechanical arm is unable to support its own weight of 411 kg, but in orbit, it is designed to handle a payload with a mass of 24.5 Mg.

Composites have become the chosen material for spacecraft antennas. The high-gain antenna system of Voyager illustrates some of the unique factors of the composites. The primary reflector on Voyager, graphite-epoxy 3.7 m in diameter, is one of the largest composite flight antennas ever fabricated. Another example is the optical bench developed for the High Energy Astronomical Observatory Mission B satellite, which required high stiffness and low CTE. The composite truss of the Application Technology satellite also used graphite-epoxy, which is 50% lighter than the equivalent aluminum structures.

The use of composites on spacecraft is growing as experience and performance data are obtained. More recent satellites contain more than 4000 composite parts. Major composite structures made from graphite-epoxy and graphite-Kevlar-epoxy include 11 antennas, antenna support truss structures, and a solar array. In satellite applications where thermal distortion of high specific stiffness is a critical factor, graphite-aluminum and graphite-magnesium are suitable composite materials.

The space shuttle was one of the first production applications of MMCs. It has 242 unidirectional boron-reinforced aluminum circular tubes that serve as mainframe and rib-truss struts, frame-stabilizing braces, and nose-landing-gear drag-brace struts, resulting in a 44% weight savings over aluminum extrusions. The boron-reinforced aluminum tubes that help support the fuselage frame of the space shuttle save 145 kg and possess lower thermal conductivity, which reduces the heat flow into the equipment of the components. The space shuttle cargo-bay doors are the largest graphite-epoxy composite structures ever built (18.3 m long) and were designed to take advantage of the low CTE of the material. The space shuttle uses about 60 pressure vessels, several of which have a Kevlar-filament overwrap with a cryoformed 301 stainless steel inner liner, as well as Inconel and titanium liners. These pressure vessels are 63.5 cm in diameter and operate at 20.7 MPa. By using composites, the tanks save an estimated 199 kg.

Other composite space shuttle components have included:

- Graphite-epoxy skins and Nomex honeycomb on the orbital maneuvering system

- Titanium I-beams and tubes reinforced with boron-reinforced epoxy in the aft thrust structures
- Filament-wound pressure vessels
- Titanium fuselage frame supports selectively reinforced with boron-reinforced epoxy
- CMC ablative tiles
- CCC nose covers

The nose cap and wing leading edges are subjected to temperatures up to about 1480 °C during reentry into Earth's atmosphere.

Other Space Applications. A less well known application of composite materials was the lunar surface drill carried to the moon during Apollo missions. Fifteen drills were connected to penetrate to a depth of 3 m. The tubular bore stem was constructed from boron-epoxy and glass-epoxy hybrid composite, and a tungsten carbide drill bit was used at the end of the stem. Drilling and impacting actions were combined in drilling through the soil and rock layers.

The current thinking about a space station is that space platform structural materials will be carried into orbit by the space shuttle and then assembled in zero-gravity space. Operational space platforms may require graphite-epoxy half-columns more than 10 m long. These columns would be fabricated by winding dry graphite fibers around a hot, tapered steel tube. Aluminum fittings would be placed on the end of the graphite, the winding would be inserted into a second steel tube, and resin would be injected into this mold set-up. The whole operation would take place in space, and the space shuttle would carry about 5000 half-columns into space in one trip.

Another project aims at fabricating composite structural beams in space itself. Carbon-reinforced polysulfone ribbon made on earth would be delivered by the space shuttle and be formed into triangular beams using a pultrusion-type process.

REFERENCES

1. *Mfg. Eng.*, April 1992, p 24
2. L.M. Sheppard, Global Outlook for the Ceramic Heat Engine, *Adv. Ceram. Mater.*, Vol 3, 1988, p 300-315
3. J.H. Gibbons, "New Structural Materials Technologies, Opportunities for Use in Advanced Ceramics and Composites," Tech Memo OTA TME 32, Congress Office of Tech Assessment, 1986
4. High Interest in Ceramics at Toyota, *Adv. Mater. Process.*, Vol 136, Sept 1989, p 21-25
5. "Advanced Ceramic Materials," U.S. Dept. of Commerce and National Research Council, 1985

6. Slow Growth for Ceramics, *Adv. Mater. Process.*, Vol 135, Jan 1989, p 29-44
7. H.E. Helms et al., "Ceramic Applications in Turbine Engines," FR 10/84, NASA CR 174715, DOE/NASA/0017-6, Contract DEN3-17; DE-A101-77CS-51040, EDR-11442, Allison Engines, Oct 1984
8. H.C. Yeh, "Improved Silicon Nitride for Advanced Heat Engines," AR 2/87, NASA CR179525, NAS3-24385, Feb 1987
9. T. Sakamoto, H. Horinouchi, and T. Maeda, "Ceramics-to-Metal Joining Technology for Gas Turbine Rotors," 89 GT 302, American Society of Mechanical Engineers, June 1989
10. L.M. Sheppard, Automotive Performance Accelerates with Ceramics, *Ceram. Bull.*, Vol 69, June 1990, p 1012-1021
11. D.C. Larsen and J.W. Adams, "Evaluation of Ceramics and Ceramic Composites for Turbine Engine Applications," Tech. Rep. 13, Contract F33615-82-C-5101, Illinois Institute of Technology Research, June 1983
12. L.J. Schioler, Heat Engine Ceramics, *Ceram. Bull.*, Feb 1985, p 268-294
13. 1987 Proc. Int. Conf. PM Aerospace Materials, *Met. Powder Rep.*, 1988
14. K. Katayama, M. Sasaki, and T. Itoh, Development of Ceramic Turbine Rotors, 88 GT 282, American Society of Manufacturing Engineers, June 1988
15. R.L. Hollman, "Design and Development of Ceramic Components," AIAA-88-3054, AIAA/ASME/SAE/ASEE 24th Joint Prop. Conf. (Boston), 11-13 July 1988
16. Japan's Rising Star, *Engineer*, 22 Mar 1984, p 30-31
17. H.E. Helms, "AGT 100 Project Summary," 88 GT 223, American Society of Manufacturing Engineers, June 1988
18. Y. Hamano et al., Development of Ceramic Components for High-Temperature Gas Turbines, *Structural Ceramics-Fracture Mechanics*, M. Doyama, S. Somiya, and R.P.H. Chang, Ed., Vol 5, Proc. MRS Int. Mtg. Advanced Materials (Tokyo), 31 May-3 June 1988, p 229-239
19. "A Competitive Assessment of the U.S. Advanced Ceramics Industry," U.S. Department of Commerce, U.S. Government Printing Office, Washington, D.C., March 1984
20. "Advanced Structural Ceramics: Technologies, Economics and Market Opportunities," GB-107, BCC Inc., Norwalk, CT, Dec 1987
21. J. Tennery, Ceramics in Engines-An International Status Report, *Ceram. Bull.*, Vol 68, Feb 1989, p 362-365
22. B.J. Busovne and J.P. Pollinger, "Development of Silicon Nitride Engine Components for Advanced Gas Turbine Applications," 90 GT 248, American Society of Manufacturing Engineers, July 1990
23. I. Stambler, U.S. and Japan Funding Automotive and Industrial Ceramic Engine R&D Programs, *Gas Turbine World*, July-Aug 1989, p 32-35
24. M. Taguchi, Applications of High-Technology Ceramics in Japanese Automobiles, *Adv. Ceram. Mater.*, Vol 2, April 1987, p 754-762
25. G.L. Boyd, "Ceramics for Gas Turbine Applications," World Materials Congr. (Chicago), 26 Sept 1988

26. T.N. Strom and P.T. Kerwin, Automotive Ceramic Turbine Research, Proc. MRS Int. Mtg. on Advanced Materials, *Proc. MRS Int. Mtg. Advanced Materials*, 31 May-3 June 1988, Materials Research Society
27. F.L. Boyd and D.M. Kreiner, "AGT 101/ATTAP Ceramic Technology Development," 89 GT 105, June 1989, American Society of Manufacturing Engineers; *J. Eng. Gas Turbines Power*, Vol 111, 1989, p 158-167
28. R.W. Ohnsorg, J.E. Funk, and H.A. Lawler, "Fabrication of ATTAP Ceramic Turbine Components," 90 GT 185, American Society of Manufacturing Engineers, June 1990
29. J.M. Schoenung, "Modeling the Fabrication Costs of Advanced Ceramic Engine Components," World Materials Congr. (Chicago), 26 Sept 1988
30. M. Patiky, Ceramic Turbines for Cars Could Wind Up on Planes, *High Technol.*, April 1987, p 56-58
31. S.J. Grisaffe, Ceramic Matrix Composites, *Adv. Mater. Process.*, Vol 137, Jan 1990, p 43-44, 93-94
32. M.D. Meiser, Development of Monolithic and Composite Ceramics at Allied-Signal Aerospace Company, *Structural Ceramics-Fracture Mechanics*, M. Doyama, S. Somiya, and R.P.H. Chang, Ed., Vol 5, Proc. MRS Int. Mtg. Advanced Materials (Tokyo), 31 May-3 June 1988, p 187-200
33. R.M. McMeeking, Toughening Mechanisms in Brittle Matrix Composites, *Structural Ceramics-Fracture Mechanics*, M. Doyama, S. Somiya, and R.P.H. Chang, Ed., Vol 5, Proc. MRS Int. Mtg. Advanced Materials (Tokyo), 31 May-3 June 1988, p 277-294
34. M.J. Koczak et al., Inorganic Composite Materials in Japan; Status and Trends, Monograph M7, Office of Naval Research, Far East Office, Nov 1989
35. A.G. Evans and D.B. Marshall, The Mechanical Behavior of Ceramic Matrix Composites, *Acta Metall.*, Vol 37, Oct 1989, p 2567-2583
36. R.J. Kerans et al., The Role of the Fiber-Matrix Interface in Ceramic Composites, *Ceram. Bull.*, Vol 68, Feb 1989, p 429-442
37. A.S. Brown, Taming Ceramic Fiber, *Aerosp. Am.*, May 1989, p 14-22
38. R. Warren, Ceramic-Matrix Composites, *Composites*, Vol 18, p 86-88
39. R.M. Laine, "Silicon Nitride Ceramic Fibers from Preceramic Polymers," Report TR 8, N00014-84C0392, Stanford Research Institute, June 1987
40. Q.H. Atwell et al., Polymer Processing, AFWAL-TR-85-4099, F33615-83-C-5006, *Fiber Technology*, Vol 1, *Composites Technology*, Vol 2, Air Force Wright Laboratories, Dec 1987
41. J.D. Buckley, *Proc. Metal Matrix, Carbon, and Ceramic Matrix Composites 1987*, CP 2482, National Aeronautics and Space Administration, Oct 1987
42. L. Kempfer, Forming the Pieces of the Ceramic Puzzle, *Mater. Eng.*, June 1990, p 23-26
43. L.L. Hench and W. Vasconcelos, Gel-Silica Science, *Ann. Rev. Mater. Sci.*, Vol 20, 1990, p 269-298
44. J.O. Stiegler and J.R. Weir, Jr., ORNL 6601, DE-AC 05-840R21400, Metals and Ceramics Division, Oak Ridge National Laboratory, Apr 1990

45. K.M. Prewo, Ceramic and Carbon Fiber Reinforce Glasses, *Proc. Metal and Ceramic Composite Processing Conf.* (Columbus, OH), 13-15 Nov 1984, Vol 1, Battelle Columbus Labs, p 139-152
46. J.J. Mecholsky, Jr., Engineering Research Needs of Advanced Ceramics and Ceramic-Matrix Composites, *Ceram. Bull.*, Vol 68, Feb 1989, p 367-375
47. D.W. Dawson, R.F. Preston, and A. Briggs, "Ceramic Fibre/Glass Matrix Composites with Outstanding Mechanical Performance," Harwell G.B., R12025, Atomic Energy Research Establishment, Oct 1986
48. J.W. Cauley and D.J. Viechnicki, Ceramics R&D at the U.S. Army Materials Technology Lab, *Ceram. Bull.*, Vol 67, Aug 1988, p 1340-1344
49. C.V. Burkland et al., CVI-Processed Ceramic Matrix Composites, *Proc. Metal and Ceramic Composite Processing Conf.* (Columbus, OH), 13-15 Nov 1984, Battelle Columbus Labs
50. P.J. Geoghegan, Ceramic Matrix Composites by Chemical Vapor Infiltration, *Proc. Metal and Ceramic Composite Processing Conf.* (Columbus, OH), 13-15 Nov 1984, Battelle Columbus Labs
51. D. Miller, "Preparation of Advanced Composite Ceramics via Novel Processing Methods," 5th Ann. ASM/ESD Advanced Composites Conf. and Exposition (Dearborn, MI), 25-28 Sept 1989
52. R.D. Veltri and F.S. Galasso, Chemical-Vapor Infiltrated Silicon Nitride, Boron Nitride and Silicon Carbide Matrix Composites, *J. Am. Ceram. Soc.*, Vol 73, July 1990, p 2137-2140
53. N.H. Tai and T.W. Chou, Modeling of Chemical Vapor Infiltration (CVI) in Al_2O_3/SiC Composites Processing, *Proc. of NASA Conf.* (Cocoa Beach, FL), 20-22 Jan 1988, J.D. Buckley, Ed., National Aeronautics and Space Administration, p 237-246
54. M.I. Mendelson, Characterization of Nextel™ 3-D Woven Fiber Structures, *Proc. of NASA Conf.* (Cocoa Beach, FL), 20-22 Jan 1988, J.D. Buckley, Ed., National Aeronautics and Space Administration, p 259-270
55. C.V. Burkland and J.M. Yang, Development of 3-D Braided Nicalon/Silicon Carbide Composite by Chemical Vapor Infiltration, *Proc. of NASA Conf.* (Cocoa Beach, FL), 20-22 Jan 1988, J.D. Buckley, Ed., National Aeronautics and Space Administration, p 271-280
56. R.M. Salinger et al., Progress in the Formation of Si/N-C Advanced Ceramic Fibers from Polymer Precursors, *Proc. of NASA Conf.* (Cocoa Beach, FL), 20-22 Jan 1988, J.D. Buckley, Ed., National Aeronautics and Space Administration, p 21-28
57. A.J. Eckel and R.C. Bradt, Thermal Expansion of Laminated, Woven, Continuous Ceramic Fiber/Chemical-Vapor-Infiltrated Silicon Carbide Matrix Composites, *J. Am. Ceram. Soc.*, Vol 73, May 1990, p 1334-1338
58. L. Leonard, Ceramic-Matrix Composites; Mettle for the Nasty Jobs, *Adv. Compos.*, July/Aug 1990, p 37-43
59. B.Q. Wang and A.V. Levy, Erosion Behavior of SiC Fiber-SiC Matrix Composites, *Wear*, Vol 138 (No. 1-2), 1990, p 125-136
60. B.H. Rabin, G.E. Korth, and R.L. Williamson, Fabrication of Titanium Carbide-Alumina Composites by Combustion, Synthesis and Subsequent Dynamic Consolidation, *J. Am. Ceram. Soc.*, Vol 73, July 1990, p 2156-2157

61. H.W. Lee and M.D. Sacks, Pressureless Sintering of SiC-Whisker-Reinforced Al_2O_3 Composites; I: Effect of Matrix Powder Surface Area, *J. Am. Ceram. Soc.*, Vol 73, July 1990, p 1884-1893
62. H.W. Lee and M.D. Sacks, Pressureless Sintering of SiC-Whisker-Reinforced Al_2O_3 Composites; II: Effects of Sintering Additives and Green Body Infiltration, *J. Am. Ceram. Soc.*, Vol 73, July 1990, p 1894-1900
63. L.C. Cook, The Repair of Aircraft Structures Using Adhesive Bonding—A Survey of Current UK Practices, *TWI J.*, Vol 392, 1989
64. R. Hall, M.D. Raizenne, and D.L. Simpson, Proposed Composite Repair Methodology for Primary Structure, *Composites*, Vol 20 (No. 5), Sept 1989, p 479-483
65. C. Ramsey, "Battle Damage Repair of Composite Structures," 34th Int. SAMPE Symp. and Exhib. (Reno, NV), 8-11 May 1989
66. T. Reinhart, "Composites Supportability in the U.S. Air Force," 34th Int. SAMPE Symp. and Exhib. (Reno, NV), 8-11 May 1989
67. M. Chichon, "Repair Adhesives: Development Criteria for Field Level Conditions," 34th Int. SAMPE Symp. and Exhib. (Reno, NV), 8-11 May 1989
68. T. Steelman, "Repair Technology for Thermoplastic Aircraft Structures (REPTAS)," 34th Int. SAMPE Symp. and Exhib. (Reno, NV), 8-11 May 1989
69. B. Purcell, "Battlefield Repair of Bonded Honeycomb Panels," 34th Int. SAMPE Symp. and Exhib. (Reno, NV), 8-11 May 1989
70. C. Hartup, "Battle Damage Repair of Thermoplastic Structure," 34th Int. SAMPE Symp. and Exhib. (Reno, NV), 8-11 May 1989
71. G. Sivy and P. Briggs, "Rapid Low-Temperature Repair System for Field Repair," 34th Int. SAMPE Symp. and Exhib. (Reno, NV), 8-11 May 1989
72. J. Connolly and D. Vannice, "Development of PMR-15 Repair Concepts," 34th Int. SAMPE Symp. and Exhib. (Reno, NV), 8-11 May 1989
73. M. Livesay, N. Smith, and E. Castenada, "Fast Simple System for Field Repair of Composite Armor and Structures," 34th Int. SAMPE Symp. and Exhib. (Reno, NV), 8-11 May 1989
74. M. Diberardino, J. Dominquez, and R. Cochran, "Bonded Field Repair Concepts Using Ambient Temperature Storable Materials," 34th Int. SAMPE Symp. and Exhib. (Reno, NV), 8-11 May 1989
75. R. Meade, "Repair of Thermoplastic Composites," 34th Int. SAMPE Symp. and Exhib. (Reno, NV), 8-11 May 1989
76. D. Bittaker, "Manufacturing Technology for Bonded Repair Procedures," 34th Int. SAMPE Symp. and Exhib. (Reno, NV), 8-11 May 1989
77. J. Mahon, "Induction Bonded Repair of Aircraft and Structure," 34th Int. SAMPE Symp. and Exhib. (Reno, NV), 8-11 May 1989
78. M.S. Sennett, "Field Repair of Composite Materials in Army Service; Planning for Future," TR-89-45, FR5/89, Materials Test Lab, Watertown, MA, May 1989
79. V.P. McConnell, In Need of Repair, *Adv. Compos.*, May-June 1989, p 60-70
80. L.K. English, Field Repair of Composite Structures, *Mech. Eng.*, Sept 1988, p 37-39

81. R.L. Coggeshall, "Boeing /NASA Composite Components Flight Service Evaluation," CR-181898, National Aeronautics and Space Administration, 1989
82. R.D. Wilson, "Advanced Composite Development for Large Transport Aircraft," 16th Congr. Int. Council of Aeronautical Sciences (ICAS) (Jerusalem, Israel), Aug 1988
83. P.M. Hergenrother and N.J. Johnston, Organic Polymeric Composites for Aerospace Applications., *Proc. Polymeric Materials: Science and Engineering*, Vol 59, 1988, p 697-701
84. Product Data Sheets. Hercules Aerospace Products Group, Narmco Materials, Inc., ICI-Fiberite Composite Structures Group, and Hexcel Structural Products, Inc.
85. H.L. Bohon and J.G. Davis, Jr., "Advanced Composites Technology-Status and Opportunities," FIBER-TEX 89 (Greenville, SC), 30 Oct-1 Nov 1989
86. N.J. Johnston and P.M. Hergenrother, "High-Performance Thermoplastics: A Review of Neat Resin and Composite Properties," TM 89104, National Aeronautics and Space Administration, Feb 1987
87. N.J. Johnston, T.W. Towel, and P.M. Hergenrother, Physical and Mechanical Properties of High Performance Thermoplastic Polymers and Their Composites, Chapter 2, *Thermoplastic Composite Materials*, Vol 7, L.A. Carlsson, Ed., Elsevier Science, Amsterdam, 1990
88. H.B. Dexter and J.G. Funk, "Impact Resistance and Interlaminar Fracture Toughness of Through-the-Thickness Reinforced Graphite/Epoxy," Paper No 86-1020, American Institute of Aeronautics and Astronautics, Apr 1986
89. H.B. Dexter and J. Maiden, "Application of Textile Material Forms to Composite Aircraft Structures," FIBER-TEX 1987 (Greenville, SC), 3-6 Nov 1987; CP-3001, National Aeronautics and Space Administration, 1988
90. D.L. Smith and H.B. Dexter, "Woven Fabric Composites with Improved Fracture Toughness and Damage Tolerance," FIBER-TEX 88 (Greenville, SC), 13-15 Sept 1988; CP3038, National Aeronautics and Space Administration, June 1989
91. M. Foley and E. Bernardon, Cost Estimation Techniques for the Design of Cost Effective Automated Systems for Manufacturing Thermoplastic Composite Structures, *Science of Advanced Materials and Process Engineering*, Vol 35, 1990, p 1321-1335
92. M.M. Schwartz, Chapter 7, 8, *Composite Materials Handbook*, 2nd ed., McGraw-Hill, 1992
93. C. Luttgeharm, "Design Considerations of Material and Process Selection for Commercial Aircraft Engine Nacelles," Tech Paper 901983, Society of Automotive Engineers, Oct 1990
94. W. Becher and M. Wadsworth, "Resin Transfer Molding (RTM): What It Is and How To Do It," Paper EM 91-218, Society of Manufacturing Engineers, March 1991

95. G.C. Sharpless, "Advancement of Braiding/Resin Transfer Molding from Commercial to Aerospace Applications," Paper EM 91-217, Society of Manufacturing Engineers, March 1991
96. S.W. Kandebo, Spaceplane Conference Highlights International Hypersonic Programs, *Aviat. Week Space Technol.*, 12 Nov 1990, p 66-68
97. T.P. Herbell and A.J. Eckel, Ceramics for Rocket Engines, *Aerosp. Eng.*, Dec 1991, p 21
98. T.D. Kasten, NASP, Expanding Space Launch Opportunity, *Aerosp. Eng.*, Nov 1991, p 15
99. A. Hirose, Y. Matsuhiro, M. Kotoh, et al., *J. Mater. Sci.*, Vol 28, 1993, p 349-355
100. A. Hirose, M. Kotoh, S. Fukumoto, et al., *Mater. Sci. Technol.*, Vol 8, 1992, p 811-815
101. S. Fukumoto, A. Hirose, and K.F. Kobayashi, Diffusion Bonding of SiC/Ti-6Al-4V Composite to Ti-6Al-4V Alloy and Fracture Behaviour of Joint, *Mater. Sci. Technol.*, Vol 9, June 1993, p 520-527
102. S. Fukumoto, A. Hirose, and K.F. Kobayashi, *Mater. Sci. Technol.*, Vol 9, June 1993, p 264-271
103. M.P. Czyz and D.J. Danko, Advanced Ceramics for Electronics, *Mach. Des.*, 25 June 1987, p 77-80
104. E.J. Stefanides, Beryllia Does It Best in Traveling Wave Tubes, *Des. News*, 23 Nov 1987, p 246-247
105. B.H. Rabin, Modified Tape Casting Method for Ceramic Joining; Application to Joining of SiC, *J. Am. Ceram. Soc.*, Vol 73, Sept 1990, p 2757-2759
106. W.C. Shumay, Jr., Copper's Expanding Role in Micro-Electronics, *Adv. Mater. Process.*, Vol 132, Dec 1987, p 54-60
107. J.C. Southern, "Ceramic Aluminas; Manufacturing, Use and Characteristics," EM89-118, Advanced Ceramics '89 (Philadelphia), 20-22 Feb 1989, Society of Mining Engineers
108. H.L. Bohon, A.J. Chapman III, and H.A. Leybold, "Ground Test Experience with Large Composite Structures for Commercial Transports," Paper B-ME-83-08-8000, National Specialists' Meeting on Composite Structures, Am. Helicopter Soc., March 1983
109. H.L. Bohon, "Composites in Today's and Tomorrow's U.S. Airliners," Airmec 1985 Conf. (Dusseldorf), Feb 1985
110. M.B. Dow and D.L. Smith, Damage-Tolerant Composite Materials Produced by Stitching Carbon Fabrics, *Int. SAMPE Tech Conf Series*, Vol 21, 1989, p 595-605
111. R. Palmer and F. Curzio, "Cost-Effective Composites Using Multi-Needle Stitching and RTM/VIM," FIBER-TEX 88 (Greenville, SC), 13-15 Sept 1988; CP-3038, National Aeronautics and Space Administration, June 1989
112. H.L. Bohon and J.G. Davis, Jr., Composites for Large Transports—Facing the Challenge, *Aerosp. Am.*, June 1984
113. F.W. Tortolano, Birth of a Space-Age Plane, *Des. News*, 4 April 1988, p 52-57

BIBLIOGRAPHY

- I.F. Gehring, Carbon Fiber Polymer Composite Materials in Weaving Machinery Engineering—Examples of Practical Applications, 15th Joint Meeting of the Aachen Textile Research Institute (Aachen), 6 Oct 1988, *Melliand Textilberichte*, Vol 70, 1989, p 413
- H.J. Gysin, Advanced Thermoplastic Matrix Composites in Mechanical Applications, *Proc. Fibre Reinforced Composites FRC '90*, Institute Mechanical Engineering, 1990, p 69-72
- I. Kuck, Development of FRP Rear Axle Components in Materials and Processing - Move into the 90's, *Proc. 20th Int. Eur. Chapter Conf. of SAMPE Materials Science Monographs*, Vol 55, Elsevier, Amsterdam, 1989, p 39-50
- E.M. Silverman and R.J. Jones, Graphite Thermoplastic Composites for Spacecraft Applications, *Proc. 33rd Int. SAMPE Symp.*, 1988, p 1418-1432
- E.M. Silverman, J.E. Sathoff, and W.C. Forbes, Design of High Stiffness and Low CTE Thermoplastic Composite Spacecraft Structures, *SAMPE J.*, Vol 25 (No. 5), Sept/Oct 1989
- E.M. Silverman, R.A. Criese, and W.C. Forbes, Property Performance of Thermoplastic Composites for Spacecraft Systems, *SAMPE J.*, Vol 25 (No. 6), Nov/Dec 1989, p 38-47
- S. Walmsley, The Manufacturing Challenge for Advanced Materials for Hotol, *Proc. PRI Conf. Automated Composites '88*, Oct 1988
- B. Wenden and A. Krawanja, APC-2: Composite Thermoplastique Haute Performance a la Conquete d'Applications Industrielles, *23 Journes Europeennes des Composites* (Paris), 6-8 April 1988
- D.F. Williams, A. McNamara, and R.M.Turner, Potential of Polyether Etherketone (PEEK) and Carbon-Fibre-Reinforced PEEK in Medical Applications, *J. Mater. Sci. Lett.*, Vol 6, 1987, p 188-190

Chapter 2

Joining Polymer Matrix Composites and Resin MatrixComposites

No other technology equals, in its potential for improving operations in all industries, the contribution from new and advanced composites.

The design of polymer matrix composite (PMC) structure has mimicked that of metal structure. However, such "black" metal design does not make the best use of polymeric composites. Designing for producibility and cost has been a trial-and-error procedure. Designs have been produced with weight savings as the prime goal and cost factors included where possible.

The "Achilles heel" of many structural designs is the joint configuration and method of joining. As advanced composite parts become larger and more complex, there is a corresponding need for improved methods of assembly. In aircraft structures, one objective is to reduce the number of metallic fasteners that add weight and increase cost in terms of assembly time. Thermoplastic composites offer design advantages because they can be joined in more ways than thermosetting materials, which are limited to fasteners and adhesive bonding.

The meltability of thermoplastics and their rapid forming consolidation cycles compare very favorably with the lengthy cure and postcure cycles typically needed for thermosets. Previous work has shown that thermoplastic composites can be joined by local consolidation of an interface, using a range of techniques. These include ultrasonic welding, induction bonding, dual resin bonding (amorphous thermoplastic film), resistance heating, and focused infrared energy. However, the joining technology developed for thermoplastic composites is still very much in its infancy. Rapid and reliable methods must be developed if thermoplastic composites are to become cost-effective. Other requirements include adaptability to various joint configurations, online inspection capability, and adaptability to automation within a production environment.

Designing for repairability and manufacturing must be done simultaneously, because their implementation may cause conflicts in the final configuration of a product. For example, non-honeycomb structures are in most instances easiest to maintain and repair, but they can cost more to manufacture than individual structures. The simplicity of repair must also be balanced by the expected structural performance.

The joining technique used on a particular composite depends on the application and the material composition. For instance, composites used in aircraft are usually joined by a combination of mechanical fasteners and adhesives, whereas those used in automobiles are often joined only with adhesives.

Adhesive Bonding

Theoretically, all composites could be adhesively bonded. However, many manufacturers avoid adhesive bonds where joints undergo large amounts of stress, so fasteners are still specified for many joints. In addition, the large size of some structures and components precludes the use of the special lay-up tooling and curing equipment needed for most adhesive applications.

The use of adhesives technology to replace or support more conventional joining methods is currently at the forefront of the industry. This is especially true for composite bonding, with particular reference to large structures and unprepared surfaces (Ref 1-4). One of the major considerations is the surface of the composite, which may have to be treated to obtain adequate initial wetting and adhesion and/or long-term durability of the adhesive-composite interface. Both thermosetting and thermoplastic matrices are affected. A second major consideration is the design of the adhesive joint, particularly when bonding anisotropic materials such as multidirectional fiber composites. For example, when lap joints are stressed, they give rise to high tensile or peeling stresses that act normal to the main fiber direction. This is parallel to the direction where the composite has relatively low strength. By careful attention to the design of the ends of typical lap joints, these peeling stresses could be minimized and the strength of the joint greatly increased.

Surface Treatment

During adhesive bonding, surface preparation replaces hole preparation as the major concern for making a successful joint. The amount of surface preparation required for a reliable adhesive bond depends on the material being bonded and the adhesive used. From a manufacturing standpoint, it is desirable to perform the least amount of surface preparation possible to guarantee a reliable bond. Certain standard actions always enhance bonding, regardless of adhesive selection. A good solvent wipe of the substrate and the fastener base is the minimum surface preparation required. Abrasion of the surfaces by means of a scuff pad, sandpaper, or grit blasting, prior to solvent wiping, provides a surface condition that is generally accepted as excellent for successful adhesive bonding (Ref 5).

In recent years, researchers have found that plasma treatment (Ref 6-12), corona discharge pretreatment (Ref 13), chromic/sulfuric acid etch and gas plasma (Ref 14), laser radiation (Ref 14), fluorine pretreatment (Ref 14), and other methods prepare composite and other substrates for adhesive bonding.

Plasma treatment consists of placing the article in a chamber that is evacuated by vacuum pump. A specific gas, such as oxygen, argon, or air, depending on the treatment desired, is flowed through the chamber and drawn out of the exhaust. The gas is then excited with radio-frequency energy by applying power to the chamber electrodes. This strips electrons from the gas molecules, forming free radicals that react with the surface to be treated and create sites for bonding. The glow associated with plasma treatment is a result of stripped electrons falling back to a lower energy state, emitting light as in neon bulbs, fluorescent bulbs, and lightning. Water, oils, and other organic materials on the surface are also attacked and broken down by free radicals. Their volatile remains are swept away by the vacuum system, leaving the surface ultraclean.

Gas plasma is a good surface preparation for the adhesive bonding of plastics because it involves no solvents, chemical fumes, disposal problems, or heat that could harm the plastic surface. The plasma reacts to a depth of 0.01 to 0.1 μm (Ref 8). The effect of plasma treatment is to reduce the contact angle (a function of surface energy) of the substrate being tested. This reduction in contact angle results in improved wetting of adhesives on the treated surface and higher bond strengths. When the substrate stands in air, contact angles begin to increase again, so adhesive materials should be applied to treated surfaces as soon as possible.

According to a study conducted by Osterndorf et al. (Ref 7), strong, durable bonds are obtained with the proper combination of substrate (Noryl, Vectra, or Torlon), surface preparation, and adhesive. The type of gas used in plasma treatment also affects lap-shear strength results. Ammonia is best with Noryl plastic, argon gas is best with Vectra plastic (Ref 9), and oxygen gas is best with Torlon plastic (Ref 10, 11).

Carbon-polyether etherketone (carbon-PEEK) composites were treated with plasma by Occhiello et al. (Ref 12) to improve adhesion with conventional epoxy adhesives. Oxidizing (oxygen, air) and inert (nitrogen, argon) mixtures were used, and the effects of treatment time and radio-frequency power were investigated. Plasma treatment introduced oxygen-containing and, in some cases, nitrogen-containing functionality at the surface of the composite and increased wettability. While untreated samples showed no pull strength and adhesive failure, plasma-treated samples reached high pull strengths and cohesive failure even at very short treatment times (<30 sec). Pull strength values were relatively insensitive to the nature of the treatment gas and plasma parameters, and aging not only did not affect adhesion but in some cases improved it. These effects suggest the occurrence of plasma-induced cross-linking of the surface layer.

Kinloch et al. (Ref 13) reported how effective corona discharge pretreatment had been in obtaining adhesion between thermoplastics using toughened epoxy adhesives. Recent advances have permitted some exciting new developments in the field of surface treatment, where an oxygen plasma treatment could be equally effective.

Research at Lockheed Aeronautical Systems Co. (Ref 14) indicates that treatment of PEEK-graphite composites with chromic/sulfuric acid etch, gas plasma, and Kevlar peel ply is good surface preparation for epoxy-adhesive bonding.

Scientists at the Ministry of Defense in Israel (Ref 14) are using ArF excimer lasers and CO_2 lasers to improve bonding of engineering plastics such as polycarbonate and polyether-imide (PEI). Irradiation dissociates some surface molecules into reactive fragments. Researchers believe that the ArF laser massively alters the surface chemistry, whereas the CO_2 laser mainly softens the thermoplastic surface.

Researchers at Air Products (Ref 14) have been investigating in recent years the pretreatment of polyethylene, polypropylene, and acetal with fluorine to improve adhesive joint strengths. They're now working jointly with fluorine-nitrogen and fluorine-oxygen-nitrogen mixtures for pretreating low-density and high-density polyethylene films. Large increases in adhesion are possible with all mixtures containing fluorine.

Designs for Joining Composites

"Successful bonding" implies consistent strength and long-term reliability. In bonding of surface-mounted fasteners, strength is derived from the bond line. To achieve consistent strength, the bond line thickness must be controlled so that repeatable joints with similar performance characteristics are achieved (Ref 15).

Although high stiffnesses and strengths can be attained for composite laminates, these characteristics are quite different from those of ordinary materials to which we generally want to fasten composite laminates. The full strength and stiffness characteristics of the laminate often cannot be transferred through the joint without a significant weight penalty. Thus, joints and other fastening devices are critical to the successful use of composite materials. The specific design of a laminate joint is much too complex to cover in detail in this book; the published state of the art is summarized in Ref 16 and 17.

Bonded versus Bolted Joints. Advanced composites introduce design problems different from those common to metal and molded-plastic construction, particularly where components must be fastened together. Because most types of composite joints, whether they are adhesively bonded or mechanically fastened, involve some cutting or machining of the strength-providing fibers, joint configurations require careful planning to minimize the possibility of failure. The planning must consider weakness in in-plane shear, transverse tension, interlaminar

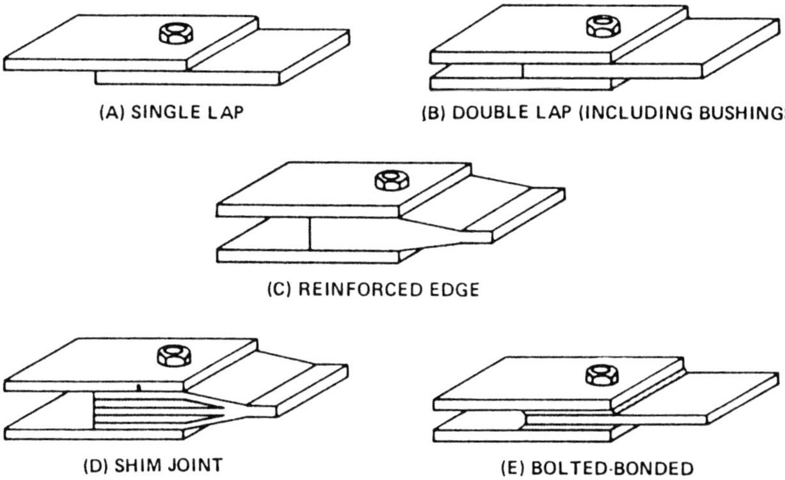

Fig. 2-1 Bolted joints. Source: Ref 17

shear, and bearing strength relative to the primary asset of a lamina, the strength and stiffness in the fiber direction.

Bolted Joints (Fig. 2-1). Because most engineered structural composites are made with thermosetting resins, they cannot be joined by welding methods, as thermoplastics and metals can. The choice for composites lies between mechanical methods and adhesive bonding. Each technique produces a joint with significant differences in production and function; each has its advantages and limitations. Mechanical joints are generally not as adversely affected by thermal cycling or humidity as bonded joints are. They permit disassembly without destruction of the substrate, and they are readily inspected for joint quality. Mechanical joints also require little or no surface preparation of the substrate and do not require white-glove and clean-room production conditions.

On the less favorable side, mechanical methods add weight and bulk to a joint, and because holes must be machined in the substrate members to accommodate bolts or screws, the members are weakened. In addition, the stress concentrations produced by mechanical fasteners can cause joint failure.

Bonded Joints (Fig. 2-2, 2-3). Performance and safety requirements of the various applications for advanced composites differ widely, and design approaches differ accordingly. For example, the importance of a failure in a carbon-fiber-composite tennis racket frame in no way approaches that of a failure in a composite helicopter rotor blade. Thus, the consequences of component failure determine the type of design data needed and the nature of testing required. The special difference that sets apart orthotropic materials, such as the advanced composites, from isotropic materials is that damage from drilling and machining makes them susceptible to interlaminar shear, delamination, and peeling.

For this reason, bonded joints are usually preferred over mechanically fastened joints for assembling advanced composite components to each other or to metal joints. More specifically, lap joints and strap joints are usually best because they

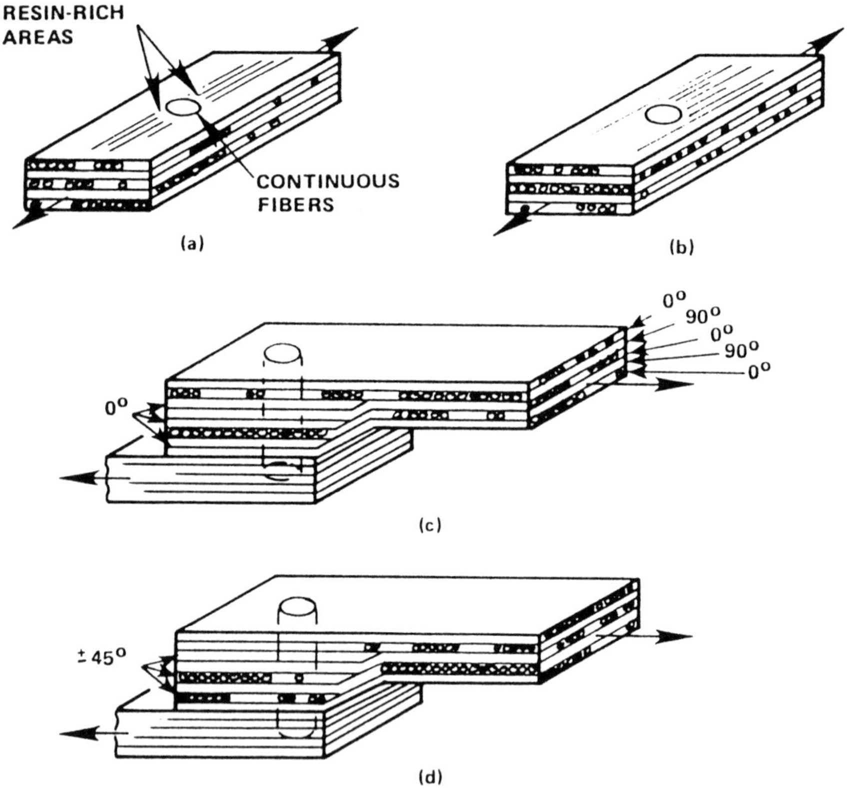

Fig. 2-2 Bonded joints. When holes are needed, bonded joints in advanced composite structures are preferred, but when strength requirements exceed those available from adhesives alone, bolts are used, often in conjunction with an adhesive. (a) Avoid molded-in holes, where continuous fiber is forced around the plug that forms the hole, leaving resin-rich areas that are relatively weak. (b) Drilled holes are better. The cross-fibers at 45° or 90° to the load maintain transverse integrity and keep cracks from forming. (c) Do not reinforce a hole or cutout with carbon fibers aligned in the same direction as the load; such reinforcement merely stiffens the joint and makes it more prone to failure. (d) A better treatment is to soften the joint by adding staggered plies of ±45° fabric. Glass fiber is better than carbon fabric here because glass has a higher strain capacity. Metal shims are also acceptable. Source: Ref 17

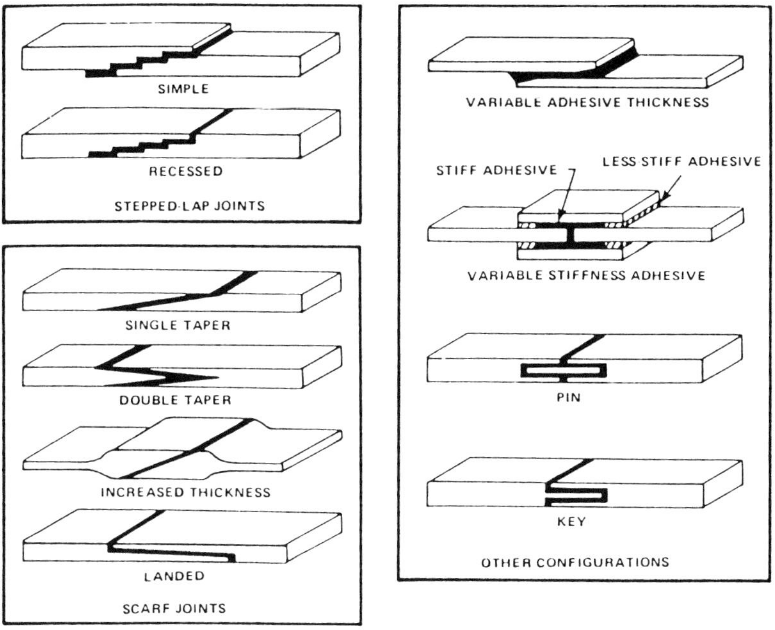

Fig. 2-3 Bonded joint configurations; scarf and other joints. Source: Ref 18

require no machining of the adherends. Advantages of adhesive-bonded joints include light weight, distribution of the load over a larger area than in mechanical joints, and elimination of the need for drilled holes that weaken the structural members. The drawbacks are that they are difficult to inspect for bond integrity, can degrade in service from temperature and humidity cycling, and do not permit disassembly without destruction of the joined members. In addition, bonded joints require rigorous cleaning and preparation of the adhering surfaces.

Bonded-bolted joints perform better than either bonded or bolted joints alone. Bonding reduces the normal tendency of a bolted joint to shear out, and bolting decreases the likelihood of a bonded joint's debonding in an interfacial shear mode.

Joint Configuration Applications. The following examples show how the criteria and theoretical and analytical data developed for adhesive bonding by design engineers are applied with the assistance of specialists in stress, weight, and materials to test joints and actual parts:

- Of the two configurations in Fig. 2-4, the honeycomb panel has the more structurally efficient geometric shape. Therefore, the stiffened panel can be

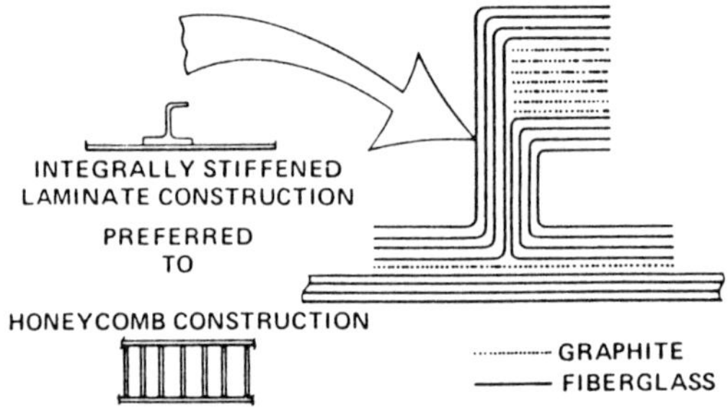

Fig. 2-4 Comparison of two joint configurations: honeycomb construction and integrally stiffened laminate construction. Source: Ref 7

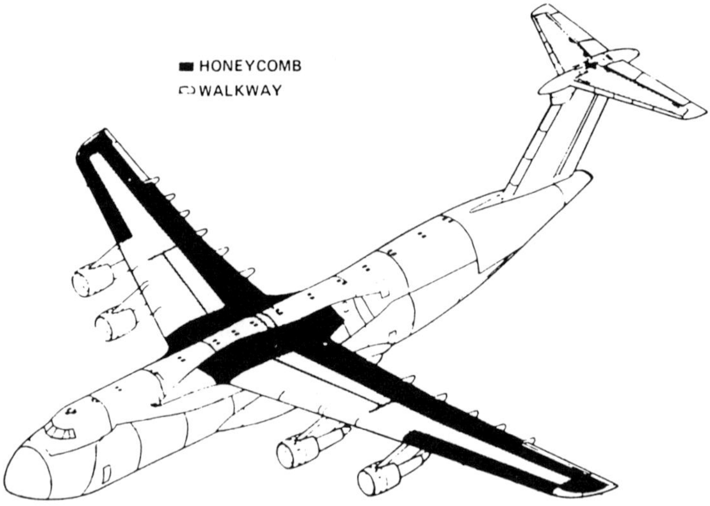

Fig. 2-5 Wing-to-fuselage aft fairings for the C-5A. Source: Ref 19

weight-competitive only if proper consideration is given to the fact that honeycomb face sheets must exceed a certain minimum thickness in order to survive the normal wear and tear of in-service conditions. For exterior aircraft surfaces, the minimum face sheet thickness varies from one aircraft to another. Thus, in applications where minimum gages must be adhered to (e.g., external fairings), an integrally stiffened laminate fairing of graphite-epoxy and glass-epoxy can be weight-competitive with commonly used fiberglass, faced honeycomb fairing. Figure 2-5 shows the wing-to-fuselage fairings that were designed from existing honeycomb construction to use integrally stiffened bonded laminates.

- A recent study of an all-composite rear fuselage for an advanced helicopter showed that Kevlar skins with graphite stiffening elements in a "skin skeleton" (Fig. 2-6) is a principal feature in achieving low-cost design. In skin-skeleton construction the skin is molded as a one-piece component, and a mating skeleton integrates frames, longerons, and stringers into a one-piece molding. The two units are then bonded together to form a complete structure.
- Figure 2-7 shows a bonded fiberglass composite main rotor blade for a commercial helicopter.
- Figure 2-8 shows an adhesive-bonded wing torque box in which boron-epoxy and graphite-epoxy composites were used.
- Figure 2-9 shows composite rudders of graphite-PI, boron-PI, and boron-epoxy for the F-4 aircraft.

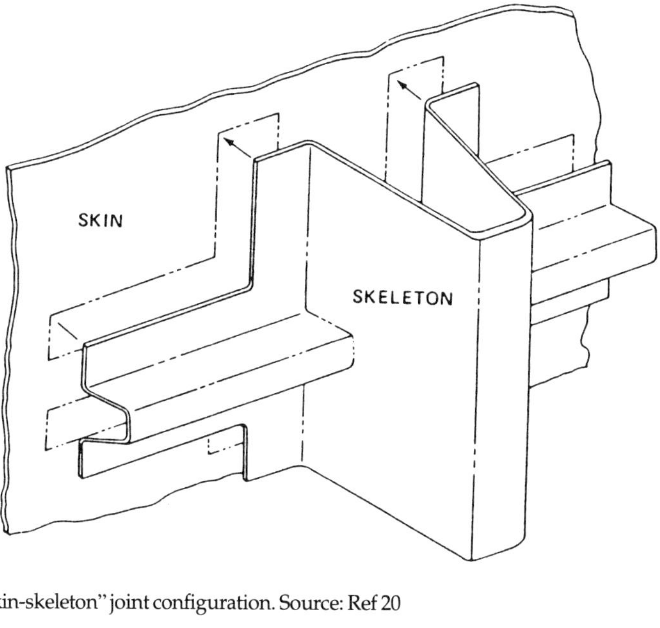

Fig. 2-6 "Skin-skeleton" joint configuration. Source: Ref 20

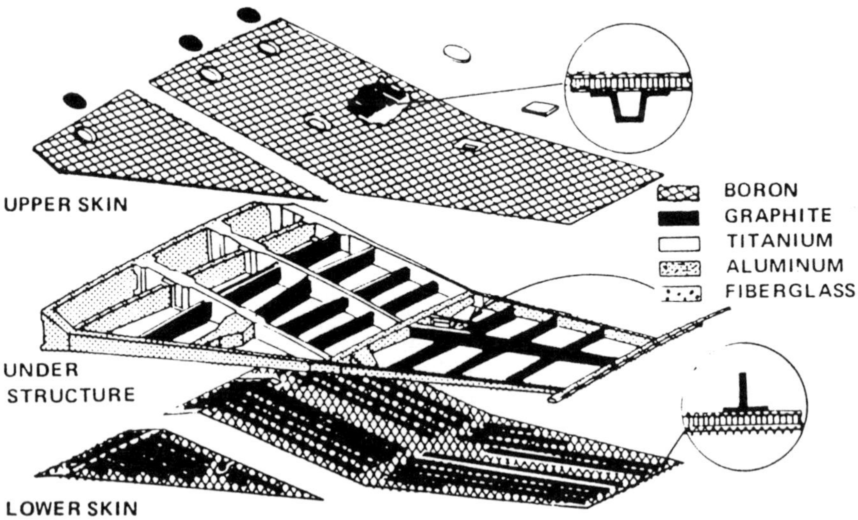

Fig. 2-7 Breakout of a fiberglass composite main rotor blade for a commercial helicopter. Source: Ref 21

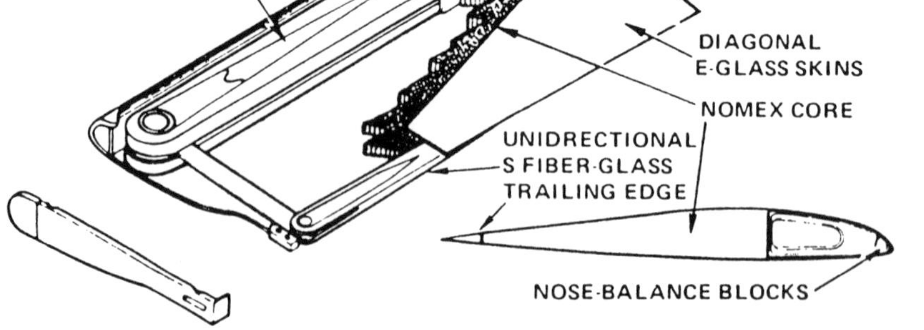

Fig. 2-8 Composite flight-test wing for the F-15 (McDonnell-Douglas). Boron-epoxy and graphite-epoxy composites were used in the adhesive-bonded wing torque box. Source: Ref 22

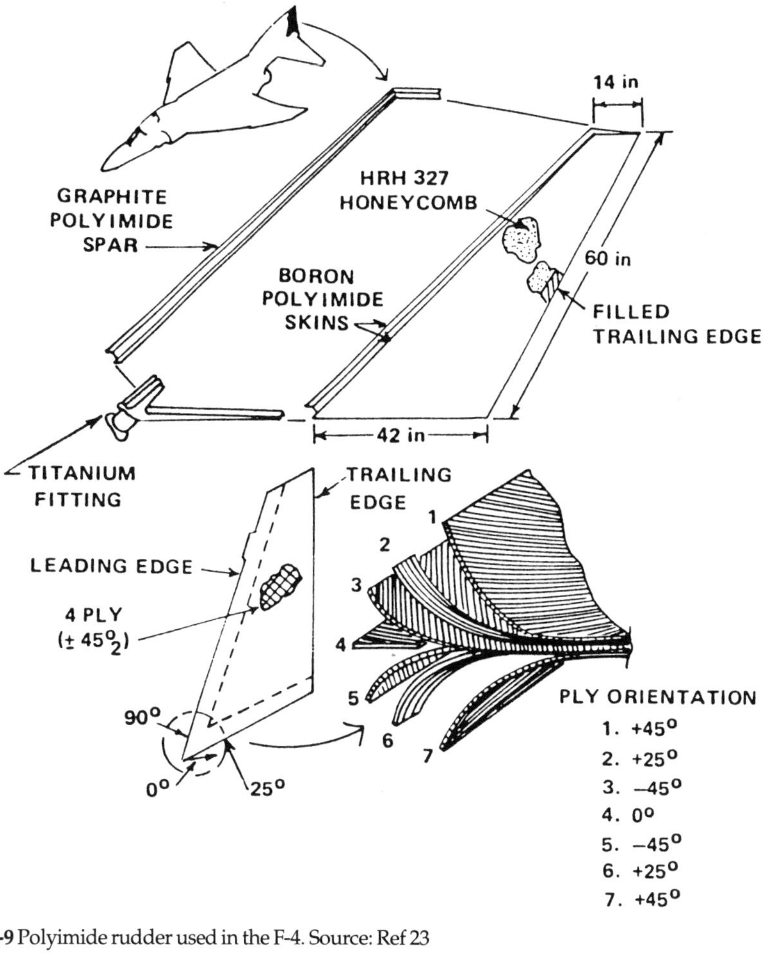

Fig. 2-9 Polyimide rudder used in the F-4. Source: Ref 23

Adhesives

A general relationship exists between the viscosity of an adhesive and the optimum bond line thickness range for that adhesive. A low-viscosity adhesive will have a thin optimum bond line thickness, and a higher-viscosity adhesive will require a thicker bond line to achieve optimum performance characteristics such as bond strength and long-term bond reliability.

Adhesive development of resin systems has intensified during the past few years. Applications for structures include engine nacelles that require long-time exposure to temperatures in the range of 150 to 300 °C, supersonic military aircraft that require moderately long exposure to temperatures of 150 to 200 °C, and missiles that require only brief exposure to temperatures up to 500 °C and above.

Because of fatigue considerations, whenever possible, it is preferable to bond rather than mechanically fasten composite structures. For this reason, the increased use of high-temperature resin-matrix composites (RMCs) has necessitated the development of comparable and equally heat-stable adhesive systems (Ref 24, 25).

Epoxy adhesives based on multifunctional resins are available that exhibit excellent strength retention at temperatures up to about 225 °C. Where long-term aging is required, epoxies are generally limited to applications requiring continuous service at temperatures no higher than 175 °C. The adherends involved are most commonly epoxy-matrix composite structures. Newly developed epoxy adhesives for use on supersonic military aircraft have excellent strength retention at temperatures up to 215 °C. Strength then drops rather sharply to 6.9 MPa at 260 °C. After 3000 h aging at 215 °C, the adhesive retains approximately 80% of its original lap shear strength. This is more than sufficient for most military aircraft. Other multifunctional epoxy adhesives are available for use in applications such as engine nacelles, where resistance to heat aging for very long periods of time is required (Fig. 2-10, 2-11).

Epoxy-phenolic adhesives rival the PI adhesives in their ability to withstand short-term exposure to extremely high temperatures. For this reason, they are admirably suited for use on missiles, and they are generally preferred to PI adhesives because of their lower cost and ease of processing. Bond pressures of 0.28 MPa and cure temperatures of 150 °C are usually adequate for most applications.

Condensation-Reaction PI Adhesives. High-temperature adhesives based on condensation-reaction PI precursors have been marketed for over 15 years. Adhesives of this type are supplied in both liquid form and as solid films. In addition, they may be filled or unfilled. Condensation-reaction PI adhesives require high cure temperatures, 260 °C or even higher, depending on the expected service temperature. Processing is further complicated by the highly volatile content of these adhesives. One processing advantage that they have over addition-reaction PI adhesives is that they can be cured under pressure at 175 °C and then postcured without pressure at higher temperatures. This permits the use of low-temperature autoclaves and conventional bagging materials. Inexpensive ovens can be used for postcuring the bonded panels.

FM34B-18, an arsenic-free compound, shows excellent strength retention after 1200 h of heat aging at 260 °C but degrades rapidly during the next 2000 h. In applications where transition metals are not involved (for example, in bonding to PI composites), FM34B-18 is as fully resistant to long-time heat aging as its predecessor, FM34.

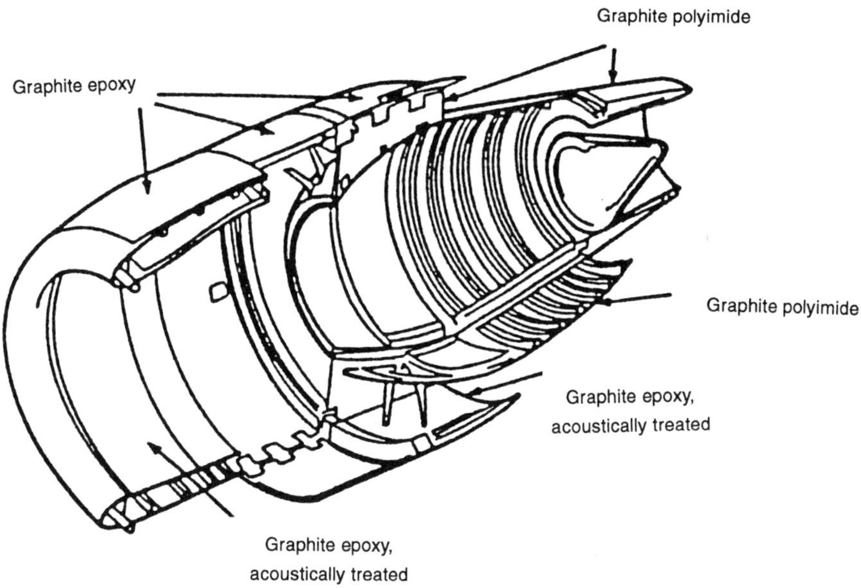

Fig. 2-10 Composite applications in an advanced-technology engine nacelle

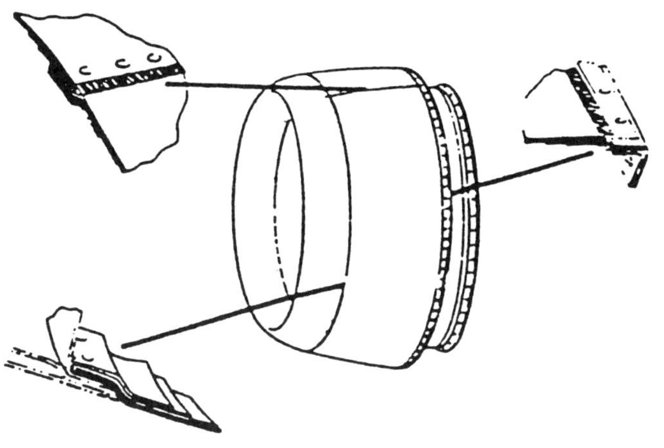

Fig. 2-11 Composite inlet outer barrel for the DC9-50

Addition-reaction PI adhesives make it possible to obtain bond lines with extremely low void contents. Adhesives of this type are generally supplied as supported film containing sufficient alcohol to impart tack and drape to the adhesive. Unlike the condensation-reaction type, addition-reaction PI adhesives remain thermoplastic after imidization and solvent removal have occurred. Another difference is that addition-reaction PI adhesives must be held under pressure throughout the cure cycle, so relatively costly high-temperature presses or autoclaves are required for fabricating panels. Composites made from addition-reaction PI adhesive are generally fabricated using 1.4 MPa to ensure low void content in the laminate.

BMI adhesives fill a niche between high-temperature epoxy and PI adhesives. Unmodified BMI resins are hard, brittle solids, and they must be plasticized in order to impart drape to the adhesive films based on them. However, unlike the PI resins, plasticization can be accomplished without resorting to the use of solvents (Ref 16, 17). Cure temperatures vary, depending on the BMI resin employed and the specific monomeric plasticizers selected. In general, good results can be obtained using a cure of 2 h at 175 °C under pressure of 0.28 MPa, followed by a 2 to 4 h postcure at 200 to 225 °C. The strength retention of BMI adhesives is surprisingly good up to about 300 °C, but in applications requiring long-term exposure to high temperatures, BMI adhesives are not expected to be durable beyond 200 to 225 °C. Almost limitless variations in formulation are possible with BMI adhesives.

Structural Adhesives. Advances in structural adhesives technology are revolutionizing methods of manufacture and assembly in virtually every modern industry, for many reasons. Structural adhesives offer significant advantages over traditional fastening methods: increased basic strength, greater durability, and remarkable resistance to water, outdoor environments, solvents, impact, corrosion, temperature extremes, and thermal cycling. Structural adhesives also impart greater "give" than mechanical fastening methods and provide a better appearance, smooth and uncluttered by the heads of rivets and bolts. They offer improved stress distribution and even insulate and protect against galvanic corrosion.

Structural adhesives are applied in almost any way imaginable: by spraying, brushing, rolling, silk screening, or as a bead, depending on the adhesive's physical properties and the design of the bond line. They offer the user the ability to increase the performance and quality of new and existing products by using advanced materials (composites), new processes, and new quality-control techniques (Ref 26, 27). Two of the new processes are rapid adhesive bonding (Ref 28, 29) and dual-resin bonding (Ref 26, 30).

Rapid Adhesive Bonding. Most adhesive bonding is performed in heated platen presses or autoclaves that have considerable thermal mass, which significantly limits the heating and cooling rates of the work. However, thermoplastic adhesives and some thermosetting adhesives do not inherently require long processing times. A few minutes to achieve a viscosity low enough to obtain flow and wetting of the adherend surfaces, plus a short time for cure for the thermosets, is sufficient. Because of the economic savings possible, it is desirable that the adhesive bonding processes for typical aerospace composite components be accomplished in minutes, and processes have been developed to bond a variety of materials and geometries, using advanced induction heating technology at kilohertz frequencies. These concepts are designated as rapid adhesive bonding for components and field repair (Ref 28). The technology combines aerospace adhesives and induction heating to produce lightweight, compact, energy-efficient prototype equipment (Ref 31).

Rapid adhesive bonding equipment is based on a self-tuning, solid-state power oscillator that can be powered from a variety of sources, feeding kilohertz power to a ferrite toroid induction heater. The toroid geometry produces a uniform, concentrated magnetic flux into the specimen or component to be bonded, causing eddy currents to flow in a ferromagnetic susceptor and/or paramagnetic adherends. These currents heat only the bond line or its vicinity. Because no large fixtures are being heated, cooling of the work is rapid: the bonded component can usually be removed from the equipment in less than 2 min. Researchers (Ref 28, 29) have also designed a machine for seam welding composites by the toroid induction method. Bonding tests have been conducted with the toroid induction welder on several pultruded composite materials, and the resulting bond strengths have been compared to those made using a conventional bonding technique (Ref 31).

Dual-Resin Bonding. A process that uses a low-melting-point analog of the high-temperature thermoplastic used in the composite matrix, dual-resin bonding lets users melt and bond the analog without melting the composite. This Thermabond process (Ref 6) uses PEEK composites that are first coated with PEI film during part consolidation, prior to joining. Because the amorphous PEI has a glass transition temperature of 210 °C, parts can now be fused together, without sophisticated tooling, at temperatures well below the PEEK melt temperature of 340 °C. For the dual-resin or thermoplastic-interlayer process to be successful, adhesion of the amorphous layer to the composite is essential. This is accomplished by selecting an amorphous polymer that is compatible with the composite matrix resin, and/or by migration of some of the fibers from the surface of the laminate into the amorphous layer.

Several aircraft companies are evaluating the dual-resin system on thermoplastic materials. Lockheed (Ref 6) reports lap shear strengths of 28 to 35 MPa with an additional layer of neat PEEK film at the interface. Joining was accomplished at 260 to 282 °C at pressures of 0.33 to 0.60 MPa. After wet conditioning, lap shear strengths still exceeded 21 MPa when tested at 149 °C. These joints survived three lifetimes of fatigue cycling before the fixtures broke. Lockheed is also using dual-resin T-stiffeners joined to ballistic test panel skins made from graphite composites.

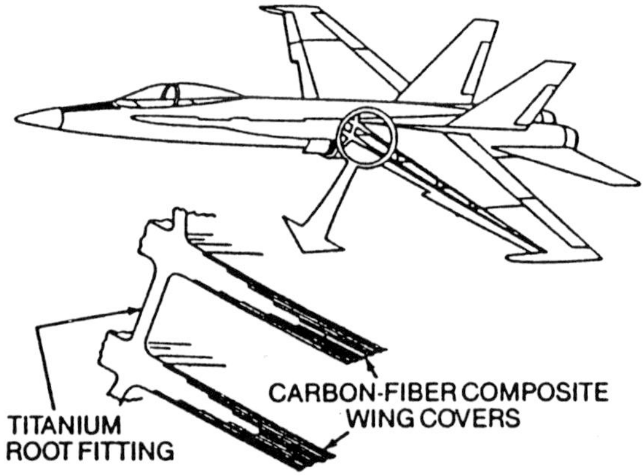

Fig. 2-12 Composite wing-to-fuselage attachment in the F-18. Source: Ref 25

In the new Advanced Tactical Fighter, there are plans to use a thermopolymer resin, dual-resin-bonded composite plus a PEI interlayer and a polyether sulfone (PES) film for skins and wings, with anticipated lap shear strengths of 35 MPa.

Applications

Figure 2-12 shows that the U.S. Navy F-18 aircraft wing is bonded to the fuselage, the ultimate in primary structure dependence on a bonded attachment. The wing joint consists of top and bottom wing cover skins that are made of carbon-fiber composite and bonded in a stepped lap-shear configuration to titanium fittings at the sides of the fuselage. The configuration removes the load gradually in a series of shear areas from the carbon composite.

Two other aircraft using composite structures are the F-15 and C-17. The F-15 has boron-epoxy composites and horizontal stabilizers with composite materials that transmit flight loads to the fuselage through primary structural adhesive-bonded joints. In addition, the composite speed brake in the aircraft is an all-bonded structure. The C-17 aircraft contains extensive applications of composites and adhesive-bonded structure. Figure 2-13 shows some of the areas proposed for bonding applications.

As manufacturers increase their use of composite materials and improve their understanding of how to design adhesively bonded structures, we will experience continued growth in the use of adhesives. This growth will be spurred by increasing demand for lighter-weight, more complex aerodynamic structures, which can best be constructed with adhesive bonding (Ref 25).

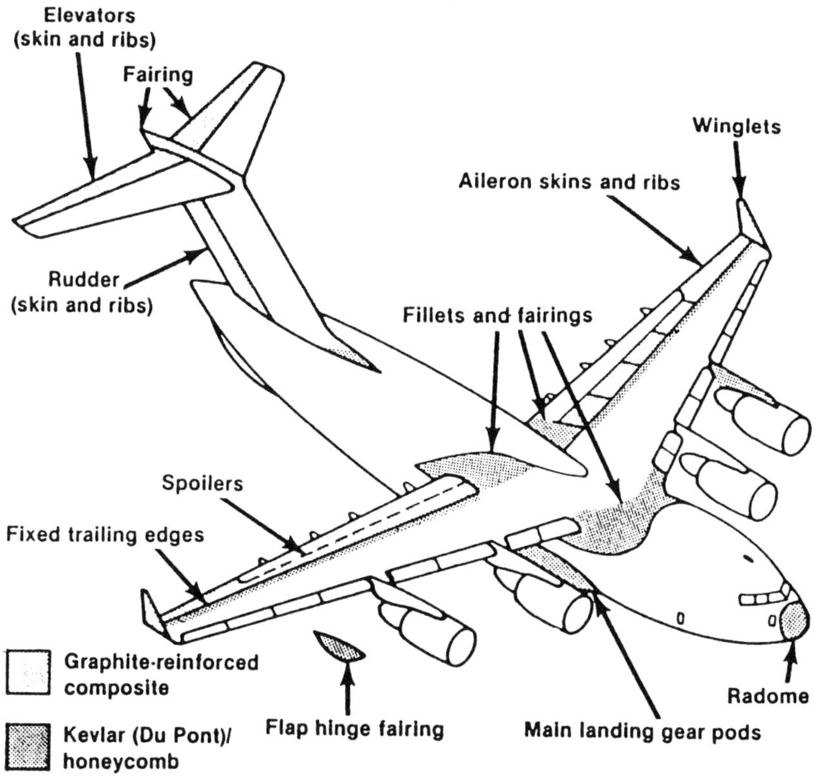

Fig. 2-13 Proposed bonding applications in the C-17. Source: Ref 32

Mechanical Fastening

By combining adhesive bonding with mechanical fastening, a partnership has been formed that resulted in the development of surface-mounted fasteners. For composite materials the benefits are twofold: the structural integrity of the material is preserved by the elimination of holes, and the structural strength is enhanced by the reinforcing effect of the baseplate, which acts as a "doubler" around the point of attachment. In the many fastening applications found in modern composite structures, the adhesive-bonded, nonpenetrating fastener is an attractive alternative method for attaching secondary systems and components.

Fastener development is evolutionary. New fastener designs are usually adaptations of existing devices, modified in response to a specific user's need. Thin sheet metal and composite materials pose special problems in fastener use, and this has given rise to entire families of inserts and blind fasteners for the aerospace and other industries using these materials (Ref 33).

Fastener Selection. Rivets, pins, two-piece bolts, and blind fasteners made of titanium, stainless steel, and aluminum are all used for composites. Several factors should be considered when specifying fasteners for composite materials:

- Differential expansion of the fastener relative to the composite
- The effect of drilling on the structural integrity of the material, as well as delamination caused by fasteners under load
- Water intrusion between the fastener and composite
- Electrical continuity of the composite and arcing between fasteners
- Possible galvanic corrosion at the composite joint
- Weight of the fastening system
- Fuel tightness of the fastening system, where applicable

Aluminum and stainless-steel fasteners expand and contract when exposed to temperature extremes, as in aircraft applications. In carbon-fiber composites, contraction and expansion of such fasteners can cause changes in clamping load. Potential clamping changes should be determined before the fastening system is chosen so joint design can be modified accordingly.

Fasteners for composites should have large heads to distribute loads over a large surface area. In this way, crushing of the composite is reduced. Fasteners should also fit closely to reduce the chances of fretting in the clearance hole. Interference fits (where the fastener is slightly larger than the hole) may cause delamination of the composite. Sleeved fasteners can limit the chances of damage in the clearance hole and still provide an interference fit. Fasteners can also be bonded in place with adhesives to reduce fretting (Ref 34-37).

Because composites have lower thickness compression strength than metals, blind fastener heads should expand more for proper fastening. The pull-up length of the fastener should precisely match the thickness of the panels being joined, to avoid local crushing. The fastener should not expand within the hole, which also causes delamination. A larger clamp area may prevent pull-through. In addition, holes must be drilled cleanly, minimizing splintering and fraying and consequent fiber separation (Ref 33).

When carbon-fiber composites are cut, fibers are exposed. These fibers can absorb water, which weakens the material and adds weight to the structure. Sealants can prevent moisture absorption in the clearance hole, but this complicates the process and adds cost. It also defeats any effort made to maintain electrical continuity between the composite fibers and the fasteners. Sleeved fasteners can provide fits that reduce water absorption and provide fuel tightness.

Lightning Strike Resistance. The use of advanced composites for airframe manufacture, particularly the joining of components, has not been without its difficulties. Among them are low damage tolerance, in-service delamination, poor production reproducibility, and uncertainties as to the compatibility of composites in replacing metal structures. One of the more interesting uncertainties is the behavior of composite structures when struck by lightning.

An inherent obstacle that engineers must overcome in using mechanically fastened composites is the dissipation of electrical energy caused by lightning strikes. It is important to dissipate such energy to avoid damage to the composite structure. Moreover, controlled energy dissipation becomes extremely critical in and around the fuel tanks, where electrical arcing can have catastrophic consequences. In composite material structures, when a lightning strike attaches to a metallic fastener, the current must be dissipated through the carbon fibers that run perpendicular to the fastener and parallel to the skin surface. For this to occur, the fastener must be in intimate contact with the carbon fibers through interference fit. In the past, fasteners installed into interference in a graphite-epoxy advanced composite structure would tend to cause damage to the structure by crushing the laminates, bending down the reinforcing fibers, and delaminating the laminate build-up (Ref 38).

A fastener has been developed, Hi-Lex (Ref 39), that is manufactured from Ti-6Al-4V for light weight and excellent corrosion compatibility with carbon-fiber-reinforced composite structures. It is generally installed in the "bare" condition, that is, without subsequent surface treatments or coatings to inhibit the conduction of electrical current. Figure 2-14 shows that the shank of Hi-Lex has a left-hand helix that requires a counterclockwise rotation to install. When the fastener head is fully seated in the joint, the driving hex shears off smoothly and flush to the surface of the fastener head. The collar is then attached with a clockwise rotation on right-hand threads. After the collar is seated, the torque is continued until the wrenching flats on the collar separate at a predetermined torque level. This results in a complete installation with the proper preload or clamp-up for beneficial joint strength and fatigue life.

In addition to lightning strike conductivity, there are other structural benefits of this unique method of obtaining interference fit. To obtain a high structural efficiency in bolted composite joints, each fastener must accept its correct share of the bond. Because composites fail at a very low level, proper load sharing between fasteners is important and is not obtainable with loose fits (Ref 40).

Special Considerations. Several factors deserve special consideration when mechanical fasteners are used with composites.

Installation of fasteners requires special care to ensure joint quality and long fatigue life. High clamp-up pressure is beneficial, especially when using clearance-fit holes. Clamping forces must be distributed over a sufficient area to avoid exceeding the composite's compressive strength. Where assembly gaps are initially present in a structure, there must be enough clamp load to close them

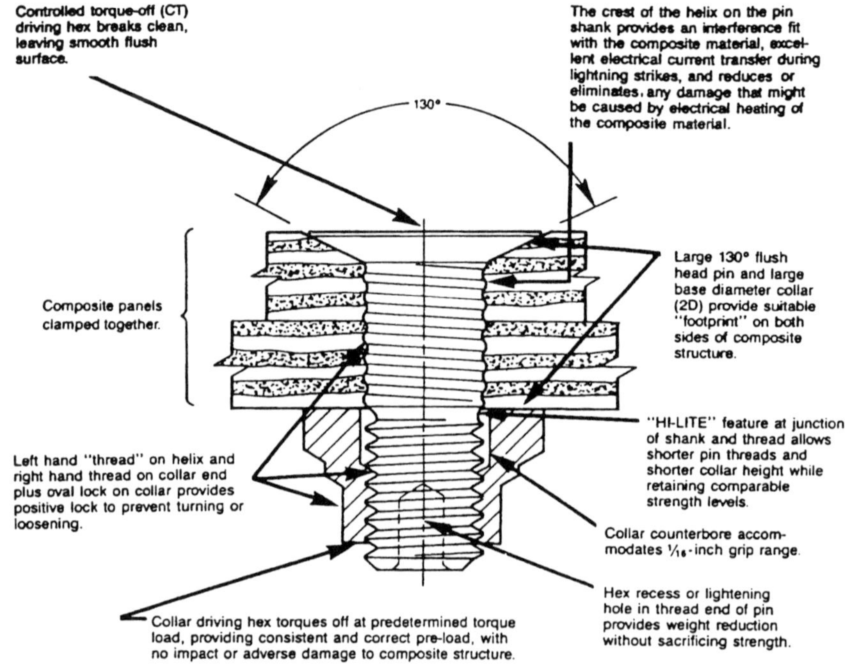

Fig. 2-14 Cross section of the Hi-Lex fastener, which can dissipate electrical energy caused by lightning strike. Source: Ref 38, 39

without damaging the structure. Assembly gaps cannot exceed the fastener's grip range, however. As a general rule they are dealt with by using temporary fasteners, some of which offer increased bearing areas for use with composites. Generally, to increase joint performance (particularly fatigue resistance), the fastener should be installed in an interference-fit hole (i.e., one that is slightly smaller than the fastener diameter), or a hole-filling fastener should be used that expands radially during installation (Ref 41).

Drilling and machining can damage composites. The number of defects, such as delamination, resin erosion, and fiber breakout, that are allowed in any structure depends on the application. For instance, because joint failure in carbon-fiber composites is caused primarily by localized bearing stress rather than overall stress, delamination is a much more serious defect than fiber breakout in a carbon-fiber composite application. Drilling techniques and the tools selected are determined by the resin, the fiber or fiber combinations in the resin, the way the fibers are configured, and the composite-metal composition of the structure (Ref 34-37).

When a composite is joined to metal, and the metal is on the blind side, the principal considerations are the desired mechanical properties of the joint, fastener material and strength, galvanic compatibility, fastener design, hole preparation, composite material thickness variations, lightning-strike resistance, and fastener installation.

Critical joint parameters in composite materials are tensile pull-through, static lap shear, and lap-shear fatigue.

Materials. Titanium alloy Ti-6Al-4V is the favorite fastener material for joining carbon-reinforced materials. It has ultimate tensile and shear strengths of 1104 and 650 MPa, respectively. For enhanced mechanical properties, Inconel 718 or A-286 corrosion-resistant steel materials can be cold-worked to develop tensile strength up to 1517 MPa and shear strength up to 758 MPa.

A carbon-reinforced composite requires that fastener materials be near carbon on the electromotive scale. An aluminum-alloy or steel-alloy fastener would quickly corrode in the presence of carbon and an electrolyte. Salt spray tests show that titanium, austenitic stainless, and certain multiphase and Inconel alloys are the best fastener materials.

A fastener must be compatible with all materials in the joint. For example, when uncoated titanium fastener is used to join a carbon-reinforced material with aluminum, the titanium fastener should be coated with aluminum for galvanic protection.

A few companies are making fasteners out of advanced composites, for use with advanced composites, and some are making a variety of fasteners for use with composite materials:

- The Huck Comp GP lock-bolt titanium fastener (Ref 42) is reported to save about 1.75 kg of weight for every 1000 fasteners.
- The Composi-Lok blind bolt (Ref 33, 42) is designed for composite-to-composite and composite-to-metallic applications where the blind bearing side is against the composite sheet. The fastener ensures structural integrity under severe vibration.
- The Live Lock fastener (Ref 42) is recommended for access panels, engine cowlings, and any removable composite sections.
- The Comp-Tite (Ref 42) fastener is primarily intended for composite-to-composite fastening applications, as in empennage and wing sections.
- A group of fasteners, Comp-Fast (Ref 42), can be customized by design and manufacture to handle various applications. They can be used to fasten composites to metals in skin fastening, structural-panel fastening, and applications with rivetless plate nuts.

Several of these fasteners are illustrated in detail in Ref 33, 41, and 42.

Moves are being made by fastener firms toward automatic installation in composites, and automation is expected to become even more prevalent. An automated system for drilling and installing Huck Comp fasteners has been reported (Ref 42) whereby 70,000 lock bolts have been used in the V-22 Osprey vertical takeoff and landing aircraft. The system is expected to have a machine cycle time of only about 9 sec. Another firm, Monogram (Ref 42), has developed end effectors for robotic installation of its blind bolts. The fairly simple pneumatic tools for the rotary-actuated blind bolts weigh about 15.5 kg, making it easier for the robotic system to maintain placement accuracy. According to Monogram, installation of blind bolts is easier to automate because blind bolts require access to only one side.

Cherry-Textron and several other firms (Ref 42, 43) have introduced an all-composite fastener. The pin-and-nut fastening system of Cherry-Textron has been produced in eight composite materials: PEEK/long carbon fiber, PEEK/short carbon fiber, Vectra/glass fiber, PEI/glass fiber, PI/glass fiber, epoxy/carbon fiber, PI/carbon fiber, and epoxy-glass.

Fusion Bonding

The chemical adhesion that holds a material together, cohesion, results from the strong valence attraction between substrates that are caused to flow together, such as when metal surfaces melt and flow together during welding. In cohesive bonding, there is an intermingling of molecules between substrates. In thermoplastics and thermoplastic composites (TPCs), the fusion process involves heating the polymer to a viscous state and physically causing polymer chains to interdiffuse, forming a weld. Welding techniques can provide bond strengths that are equal to those of the parent polymer.

Development of advanced joining technology will become essential as the demand for TPCs and engineering thermoplastics continues to grow, and as composites are used for increasingly complex parts. Figure 2-15 shows the three categories of thermoplastic welding techniques, and Fig. 2-16 compares some of the techniques in terms of tensile shear strength. All of these techniques deal with the basic problem of localizing heating to the bond line so that part configuration and structural integrity are maintained in the process. Each category—thermal welding, friction (mechanical) welding, and electromagnetic welding—is described in its own section below. Table 2-1 provides a capsule assessment of each technique, and, along with Table 2-2, compares fusion bonding techniques to adhesive bonding.

Thermal Welding

In thermal welding, heat is applied directly to the individual surfaces to be joined. The surfaces are then quickly brought into contact, held under pressure, and allowed to cool, forming a bond. Methods used to heat the surfaces include heated tools and platens, hot gas, focused infrared energy, and laser beams.

Focused infrared energy welding (FIRE), a novel noncontact fusion bonding technique, has been introduced to industry in a semicommercial phase. It is said to be particularly suited for parts with problematic geometries and for high-performance TPCs, such as carbon-fiber-reinforced PEEK or polyphenylene sulfide (PPS) composites. It is targeted for applications that require joints as strong as the composite itself.

While reinforcing fibers in a matrix dramatically upgrade the physical properties of a composite, they often make welding difficult because there is less resin available to melt and reconsolidate into a fused joint. For example, TPC laminates

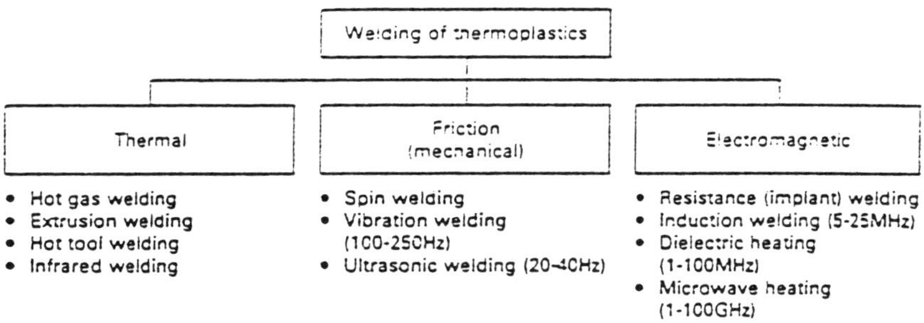

Fig. 2-15 The three categories of thermoplastic welding techniques

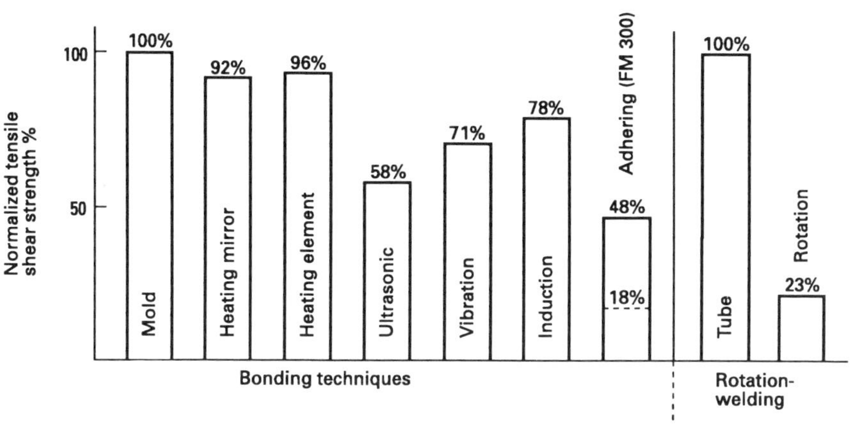

Fig. 2-16 Comparison of the normalized tensile yield strength of bonds produced by selected techniques. Source: Ref 56

contain up to 70 wt% reinforcement. Reducing the amount of reinforcement improves weldability, but at the expense of composite strength. Also, advanced thermoplastics used as matrix resins in TPCs have narrower "processing windows" and must be processed at higher temperatures than general-purpose engineering resins. This demands precise control of heating time and temperature to produce a strong weld joint. Methods such as induction, resistance, and ultrasonic

Table 2-1 Assessments of current methods for joining thermoplastic composites

Technique	Assessment
Resistance welding	Ideal for short bond lines in nonconductive composites. Requires conductive metal or carbon strips that remain in the joint after the application of resistance current; this may weaken the bond by producing a discontinuity in the integrity of the joint. A drawback of conductive thermoplastic composites is that the resistance heating element must be insulated from the conductive constituents of the composite. Ohm's law restricts the length or width of bond line in resistance welding.
Ultrasonic bonding	Ideal for spot welding in thicknesses less than 1 mm (0.040 in.) where disruption of the fiber is not a critical factor. Generally effective in layers of less than 1 mm (0.040 in.) and suitable for short bond lines under 254 mm (10 in.) long; high-frequency vibrating of the materials may dislocate and abrade the fiber structure of the composite.
Vibrational bonding	Ideal for relatively small, flat parts and some three-dimensional parts. Requires the rapid frictional abrading of the two surfaces to be bonded, which may disrupt the fiber embedded in the matrix. Generally successful with smaller parts.
Induction bonding	Ideal for nonconductive composites and where introduction of a susceptor as a permanent bond line element is acceptable. Requires the use of ferrous metal powders or particles that are left in the joint area of the interface.
High-frequency welding	Not suitable for conductive composites because high-frequency welding uses the material as a buffer. If the reinforcing fibers are dielectrically conductive, then transfer of heat to the interior of the composite may result in overheating, melting, thinning, and weakening.
Traditional infrared	With traditional approaches to the use of infrared, electromagnetic energy is absorbed and dissipated by the entire body of material. The bonding surface, when exposed to so much unfocused radiant energy, melts, deforms, and weakens. When removed from the heat the material cools before it can be brought together.
Heated air	Tends to be limited in bond line area, and sometimes ambient air conditions make temperature control difficult.
Hot plate	Ideal for lower-temperature thermoplastics. High-temperature thermoplastics require longer heating cycles during which heated metal is in physical contact with the resin of adhered interfaces. This often results in disruption of the material and difficulty in controlling welding temperatures.
Mechanical fasteners	Often unsatisfactory because the integrity of the composite sheet is interrupted by the puncture necessary to affix the fasteners; mechanical joints often leak and create weakness of the joints.
Hot-melt, room-cure adhesives	Difficult to apply uniformly and, because of the difference in composition and lack of molecular continuity, create fracture lines. Fast cooling time of hot-melt adhesives may create difficulties.

Source: Ref 44

Table 2-2 Comparative advantages of six joining techniques

Method	Reproducibility	Adaptable to various joint configurations	Suitable for small- and large-area bonding	Minimal surface preparation	Minimal equipment use	Adaptable to automation	On-line inspection capability	Rework-ability	Production environment
Adhesive bonding	10	10	10	2	9	Yes	10	No	Yes
Fasteners plus adhesive	10	10	10	1	8	Yes	10	No	Yes
Resistance heating	5	8	8	10	7	Yes	5	Yes	No
Ultrasonic welding	5	5	5	10	5	Yes	10	Yes	No
Focused infrared energy welding	10	6	4	10	5	Yes	5	Yes	Yes
Amorphous thermoplastic film	10	10	10	10	10	Yes	10	Yes	Yes

0 = lowest; 10 = highest. Source: Ref 64

welding do not lend themselves to exact temperature control because parts are clamped before heat is applied. As a result, joint temperature can't easily be measured to determine precisely when the joint has melted. Often, thermocouples must be embedded in the joint area, complicating the joining process and potentially degrading the bond.

The FIRE technique overcomes these problems by using reciprocating, focused infrared heaters that progressively apply a precise amount of intense heat to bond lines on two separated bond surfaces, which are pressed together after heating to produce a strong welded joint (Fig. 2-17). During reciprocation, the surface temperatures of both bond faces increase with each stroke. However, because the infrared beam is moving, it is not in any one position long enough to burn the resin. This also allows the surface temperature to increase to the melt point without increasing the internal temperature of the composite to the point that the matrix would be damaged by slippage or fiber separation.

Temperature sensors can be positioned to directly monitor temperature at the bond lines and to regulate the power applied to the infrared heaters. The heaters reciprocate inside the press but do not contact the parts. As a result, there is no distortion of the bond lines or part surfaces and no adhesion of melted material to the heaters. The temperatures of both surfaces can be controlled individually during welding, which permits assembly of thick and thin parts as well as thin-wall parts. The temperature of the beams can be accurately adjusted during the

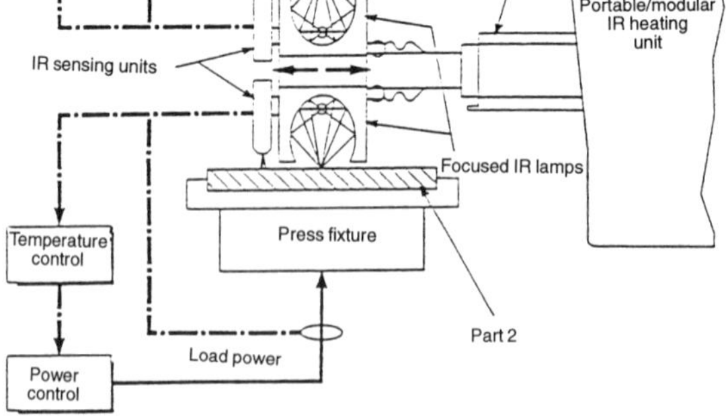

Fig. 2-17 Focused infrared energy (FIRE) welding joins thermoplastics by moving a pair of linear, focused infrared (IR) lamps back and forth along the bond lines of the material surfaces that are to be joined.

heating cycle to bring the temperature of disparate welding surfaces to the fusion temperature simultaneously.

FIRE can join any combination of straight or circular flat parts. The process can bond curved lines for applications such as fuselages or three-dimensional parts, and it also accommodates continuous welds. A wide range of materials can be welded, from very rigid, high-temperature advanced composites to low-modulus elastomers. Strong welds have been produced in glass-, carbon-, and aramid-fiber-reinforced PEEK, PEI, thermoplastic PI, PPS, polyamide-imide, polysulfones, and liquid-crystal polymers.

The reciprocation zone controls bond line width, and wide bond lines are possible by using multiple lamp fixtures placed side-by-side. For example, it's possible to heat a bond line area 51 cm wide by 3.75 m long by using four sets of focused infrared lamps, 1.2 m long and spaced approximately 10 cm apart, back-to-back.

Good wetting of a clean, release-agent-free bond line is essential for successful use of FIRE with advanced composites. Cleaning solutions of plasma treatment are preferred to sanding or abrading the surface, which may remove resin in the two facing adherend plies and loosen the fibers. A fiber-free film of resin tape should be applied to one part surface, but resin-rich surfaces on the composites to be joined may eliminate the need for this step. A benefit of FIRE is that the high temperature of the reciprocating beam (up to 650 °C) burns off surface contaminants that may be trapped in asperities of the interface layers.

Hot gas welding, also called hot air welding, makes use of a stream of heated gas or air directed at a filler rod and the joint area (Ref 44). The filler rod is then gently pushed or fed into the joint area, causing the fused polymers (TPCs only) to make contact. The equipment can be manual or semiautomatic. Normally, the filler material is a solid rod that is forced under light pressure into the joint. The filler rod is not melted in the gas stream, but it is sufficiently softened to allow it to fuse to the parent material (Ref 45-48). For best results, the filler rod should have the same composition as the substrate. Usually, only the surface is fused.

Gas temperatures will depend on the polymer, but they typically fall in the range of 200 to 300 °C, and gas flow rates range from 15 to 60 L/min. The gases used include nitrogen, air, carbon dioxide, hydrogen, and oxygen. Compressed air is popular because it is economical and satisfactory for many purposes. For smaller, manual welding operations, the results depend on the skill of the operator (Ref 49, 50).

Hot gas welding is used mostly for butt and fillet welding, but also for lap welding of thin sheet. In lap and fillet welding, no joint preparation is needed. In fillet welding the filler is fed directly into the 90° angle of the fillet, and for lap welding no filler is used (Ref 51).

Hot tool welding, also called hot plate welding, is a tolerant process capable of providing weld strengths equal to those of the parent polymer. It is perhaps the simplest welding technique used with plastics. It's popular in mass production and is used for large structures.

In some operations, a heated tool is simply used to press like materials together until melting occurs at the joint. Joints may also be heated by hot dies. The dies are removed and the polymers are quickly brought together while in the molten phase. Nearly all polymers may be made heat sealable by applying a layer or coating of material that is heat sealable.

Hot plate welding of large components with wall thicknesses above 25 mm currently presents particular challenges (Ref 51). The Welding Institute has designed and built a test-bed machine that has an electrically heated hot plate and hydraulically applied axial force in common with commercial equipment, but whose parameters can be varied over wide ranges. The maximum hot plate temperature is 650 °C, and force is adjustable from 1 to 150 kN, so all current thermoplastic high-temperature composites can be welded in sizes up to the dimensions of the hot plate (650 × 650 mm). Typical areas of current research include joining large-diameter (up to 630 mm) advanced TPC parts for aerospace use.

Friction Welding

In friction (mechanical) welding, parts are first put in contact with each other and then heat is applied, concentrated at the bond line. Examples include spin, vibration, rotational, and ultrasonic welding, all of which have been evaluated with advanced TPCs (Ref 6).

Spin welding uses the heat from friction to melt substrates together. Circular thermoplastic parts are bonded by this method. Part joints or filler rods may be spun to create the weld joint. Once the melt cools, the bond is completed. Some major welding parameters must be accounted for during the fabrication process. Rotational speeds range from 1 to 20 m/sec, friction pressures from 80 to 150 kPa, forge pressures from 100 to 300 kPa, and weld times from 1 to 20 sec.

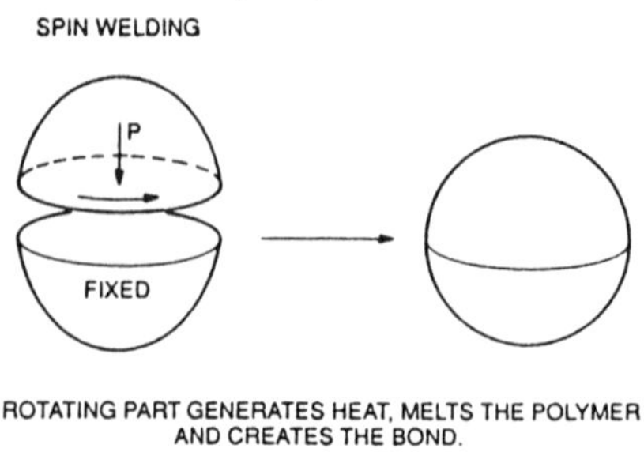

Fig. 2-18 A schematic of spin welding. Source: Ref 52

Spin welding has been performed with a drill press, but better-quality welds can be obtained with units that impart a controlled amount of energy to the spinning part. Usually this is done by fixturing one part in a flywheel that is spun to a preset energy level (Fig. 2-18). On contact with the fixed part, the energy is dissipated at the interface. Other designs use driven fixtures to impart fixed amounts of energy (Ref 52).

The advantages of spin welding are high weld quality, simplicity, speed, and reproducibility. In most cases, little surface preparation is necessary. The main disadvantage is that, in its simplest form, spin welding is suitable only for applications where at least one of the components is circular and angular alignment is not required. Orbital welding can be used in the latter case, but the process is more complex. Most TPCs can be bonded, including dissimilar polymers with compatible melt temperatures (Ref 53).

Vibration welding, also called linear friction welding, was first described in the mid-1970s and has since been increasingly often used for welding various mass-produced thermoplastic components. Frictional heat is generated by relative linear movement between the two thermoplastics to be welded, which are held under pressure. Once molten material has been generated at the joint interface, vibration is stopped, the parts are aligned, and the weld consolidates on cooling.

The process is rapid (weld times of 1 to 5 sec are generally used), the vibration is typically 100 to 240 Hz at 1 to 5 mm amplitude, and pressures are similar to those used in rotational and ultrasonic welding (1 to 4 MPa). It is similar to spin welding, except that the motion is linear rather than rotational (Ref 52, 54). The main advantage of vibration welding is its ability to weld large, complex linear joints at high production rates. Other advantages include simplicity of tooling, the ability to weld a number of components simultaneously, and the ability to weld almost all thermoplastic materials. It is more difficult to apply to continuous fiber materials because of the likelihood of fiber displacement.

Vibration welding has been tested on APC-2*, a graphite-reinforced TPC. The effects of weld times between 1 to 6 sec were investigated with weld pressures in the range of 1 to 2.5 MPa and vibration amplitudes between 2 to 3 mm. The direction of vibration amplitude was parallel to the length of the specimen. Using low pressure and low vibration amplitude resulted in little or no bonding of the surfaces. Increasing either parameter resulted in higher-strength welds, with a maximum of 17 MPa achieved (35% of parent material). At all welding conditions, the weld strength tended to increase with weld time, but the degree of flash and fiber displacement also increased. This distortion/displacement is a major disadvantage of the process when used on APC-2 (Ref 52, 55). Vibration

* APC-2 (usually expressed as APC-2/PEEK) has been made as a prepreg tape and has subsequently been made into laminates. The carbon fiber in APC-2 has usually been AS4 from Hercules. APC-2/PEI has also been produced (where PEI has been a film).

welding tests on PEEK composite laminates have realized lap shear strengths of 39.9 MPa (Ref 52, 54, 55).

Rotational welding is suitable for bonding rotationally symmetrical formed parts. One part rotates, producing heat, while the second part is held in place. After rotation is stopped the weld "freezes." The rotational speed and friction time have an important effect. A conical surface weld is chosen to maintain the favorable properties of continuous reinforcement.

Ultrasonic welding uses high-frequency (i.e., ultrasonic) mechanical vibrations for joining. The parts to be assembled are held together under pressure and are then subjected to ultrasonic vibrations (usually 20 to 40 kHz) at right angles to the contact area, so that longitudinal vibrations are transmitted through the component. The consequent alternating high-frequency stresses produce heat in the material, and if the components are properly designed this heat can be selectively generated at the joint interface. Heat generation is by a combination of surface and intermolecular friction. Welding time is normally 3 to 4 sec, with parts held in place to cool for a total cycle time of 10 to 15 sec.

The ability to weld a component using ultrasonics is governed by the design of welding equipment, the physical properties of the material to be welded, the design of the components, and the welding parameters (Fig. 2-19). Ultrasonic welding has been done successfully with graphite-PEEK composites, and lap shear strengths of the APC-2 material have been virtually equal to the strength exhibited in compression-molded laminates (78.4 MPa). For continuous-fiber composite materials, the main barrier to the use of ultrasonic welding is the difficulty of providing ultrasonic energy directors on sheet components and the consequent risk of fiber disruption at the interface under the high deformation necessary to obtain a satisfactory bond (Ref 57-60).

The size and power of the welder limit the area that can be bonded in one operation. Large parts can be joined by sequential or scan welding. In sequential welding, a bond is made, the part is indexed, and adjoining sections are welded until the part is fully joined. In scan welding, parts travel under the welder at a constant speed until bonding is complete. In both cases, lap shear strengths have been realized that are equal to 80% of the compression-molded control laminates.

Electromagnetic Welding

Electromagnetic welding uses micrometer-sized particles of iron oxide, stainless steel, ceramic, ferrite, and graphite that respond to the radio-frequency (3 to 40 MHz) magnetic field. These opaque powders or inserts must be molded in the polymer matrix, and they remain in the final weld. As the high-frequency field passes through the magnetic materials, they are induced to become hot. The surrounding polymer melts, forming the bond. The induction coil must be located as close to the joint as possible if rapid bonds are to be made, and nonmetallic

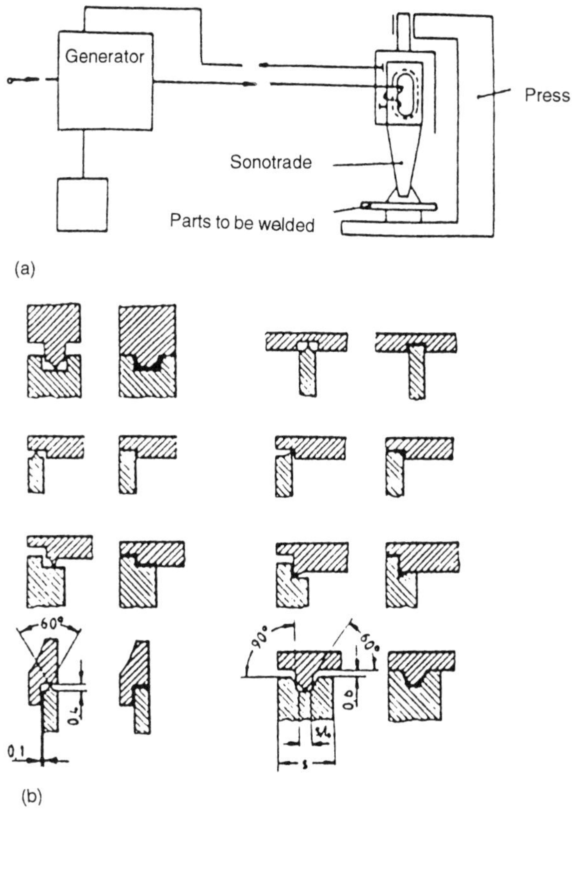

Fig. 2-19 Ultrasonic welding. (a) Apparatus. (b) Design of seams

tooling must be used for alignment. "Sandwiches," nonthermoplastic-coated composites, and reinforced, coextruded layers can be bonded. Some are bonded by incorporating magnetic particles in hot melts, liquid adhesives, or films at the joint area. Lap shear values of 26 MPa have been obtained (Ref 61). Electromagnetic welding is quick and efficient, and industry is becoming increasingly aware of its potential, especially in joining TPCs.

Induction welding is based on the principle that a magnetic field passing through a conductor generates eddy currents that dissipate in the form of heat. If the conductor is also magnetic (graphite fibers, for instance, are conductive but not magnetic), hysteresis losses also dissipate as heat. In practice, a thin layer of metal-filled adhesive or a metal screen laminated with a thermoplastic film is placed at the joint line. This metal susceptor localizes heat in the joint area once the magnetic field is applied. Lap shear strengths of 42 to 49 MPa have been reported for graphite-PEEK laminates welded in this fashion.

For application to plastics that have neither high conductivity nor high permeability, radio-frequency fields are generally applied to tapes of thermoplastic filled with iron oxide (Ref 62) or metal particles that are heated by the field. These tapes are placed at the interface, and upon application of a suitable radio-frequency field the particles cause the thermoplastic to fuse and create a weld. Coils made from copper tubing are connected to the output from the induction generator, and most of the field is generated in the coil. Coil design is thus an important parameter for induction heating. By incorporating empirical and computer modeling techniques, it is possible to optimize the work-coil design to achieve maximum eddy current generation in the vicinity of the weld.

Some PEEK, PPS, and PEI composites of graphite without metal susceptors have been successfully induction-welded (Ref 18) with lap shear strengths of 38.5 to 48.3 MPa by using APC-2 laminates and PEEK film at the interface. Some graphite-PPS composites have been welded with lap shear strengths of 23.8 to 30.8 MPa, whereas a graphite-PEI composite has been joined with lap shear strengths of 31.5 to 14.3 MPa.

A new 5 kW induction welder developed for joining thermoplastics is said to significantly reduce cycle times and to have the ability to simultaneously weld large parts and multiple bond lines. The unit features the process of thermoplastic welding, which uses electromagnetic induction welding and special material to produce structural, hermetic, or pressure-tight welds with a reportedly very high degree of reliability.

Induction welding can use the amorphous bonding technique (Ref 63), which involves making a joint by melting a layer of the amorphous polymer PEI at the weld line. (The same principle is used in resistance welding, discussed below.) PEI has a melting temperature of approximately 270 °C, whereas PEEK has a melting temperature of 345 °C, and this difference allows a joint to be made by melting the PEI while leaving the PEEK composite still fully consolidated. Induction welds using the amorphous bonding technique have achieved lap shear strengths of 40 MPa. A refinement of this technique uses ferromagnetic particles (usually iron oxide) dispersed in a thermoplastic matrix that is preplaced along the joint in the form of a paste or tape. Welding times are short, up to 20 sec for the largest components. The process can be used to join almost all thermoplastics, but the insert limits the joint strength.

Resistance welding, also called resistance implant welding, is based on the principle of trapping a metal insert between two parts to be joined and then heating the insert by induction or resistance heating, so that the plastic material around the implant fuses to form a joint. The heat causes the surrounding plastic material to melt, and the weld is effected by subsequent cooling under pressure. The conductive implant remains within the joint and as such affects the final strength of the weld.

Prepreg carbon fiber tape, specifically unidirectional fibers set in PEEK, can be used as a compatible implant material; a metallic implant would introduce foreign material into the thermoplastic joint. Welds of up to 127 by 25 mm can be achieved with a lap shear strength of up to 50 MPa. Because welding times are short, up to 20 sec for the largest components, resistance heating offers other advantages as a bonding technique for TPCs. It requires little surface preparation (sanding of surfaces), costs are low, and the process is relatively simple and fast. Joints can be consolidated in minutes at room temperature, compared to hours at 177 °C for many epoxy adhesives (Ref 52, 54, 64-68).

Fusion bonding of APC-2 by resistance welding and compression molding has been investigated (Ref 66-68). Lap shear strength from resistance-welded samples rapidly increased and asymptotically approached the compression-molded baseline in much less processing time. Resistance-welded APC-2 with PEEK film yielded a maximum lap shear strength of 44.8 MPa, with fiber motion occurring at the joint. The APC-2/PEI resistance-welded specimens achieved a maximum strength of 35.9 MPa, with no fiber motion detected. The baseline for APC-2/PEI was determined by isothermal processing (1200 sec at 293 °C) to be 43.6 MPa.

Dielectric welding, also called radio-frequency welding, is the most widely used of the electromagnetic processes. The process uses a radio transmitter, usually operating at 27 MHz, connected to electrodes that hold the sheets together, rather like the electrodes in resistance welding of metals. The equipment must be shielded to prevent transmission of its output.

Other Welding Techniques

Heated Press Joining. Tests by Cantwell et al. (Ref 69) were undertaken to examine the possibility of joining carbon-fiber-reinforced thermoplastic at temperatures below the recommended processing temperature of 380 °C. The results showed that by employing a sufficiently long contact time, it is possible to obtain very high joint strengths at temperatures just above the melting temperature of the PEEK matrix. This heated press technique thermally joined carbon-fiber-reinforced PEEK with lap-shear strengths in excess of 50 MPa. Investigation of the variation of joint strength with welding temperature showed that with a suitable choice of contact time, it is possible to join carbon fiber/PEEK successfully at 350 °C. Below this temperature the degree of molecular interdiffusion appeared to be limited, and the resulting average lap-shear strengths were low.

Polymer-coated material (PCM) welding produces joints that are said to be as strong as those achievable using adhesives, but it does so in less than a minute and by an automatic, continuous process. The technique will find broad use in the automotive and aerospace industries. It is predicted that designs based on aluminum alloy space frames with thermoplastic body panels could soon be mass produced using the PCM method. Similarly, TPCs are already finding use on civil aircraft, mainly in non-load-bearing applications. As confidence in the performance of these materials grows, their use will probably be extended to primary load-bearing structures such as ribs and spars. These applications may also demand technologies for joining plastics to metals. Examples include aluminum hinges and fittings on the doors of baggage compartments and load-bearing joints between aluminum struts and TPC wing skins.

In order to test the PCM technique, the Welding Institute (Ref 52, 61, 70, 71) used PEEK reinforced with 61% continuous carbon fibers supplied as unidirectional prepreg tape at a thickness of 0.125 mm. Sixteen layers of tape were used to form a laminate structure 2 mm thick. In addition, a layer of amorphous PEI was comolded onto one surface. A grade of aluminum alloy deemed suitable for aerostructure fabrication was also selected. A thin thermoplastic layer was applied to the surface of the pretreated aluminum alloy to facilitate the joining to the PEEK composite.

Resistance implant welding was employed because it is relatively simple to control and guarantees maximum heat input at the joint line. The implant used was a piece of unidirectional carbon-fiber prepreg tape. PEI and PES films were introduced around the implant to provide more thermoplastics in the weld. These materials are compatible with PEEK and with the thermoplastic coating on the aluminum alloy, and they melt at a lower glass transition temperature than PEEK. During welding, the glass transition temperature of PES (225 °C) and PEI (230 °C) were both exceeded well before the PEEK began to melt (at 345 °C). Thus, the added PEI and PES films, the thermoplastic coating on the metal, and the comolded PEI on the surface of the PEEK all melted together to form the weld before the PEEK was damaged by heat (Fig. 2-20). Welds were achieved by applying direct current through the carbon-fiber prepreg implant. A three-stage welding cycle was applied. The combination of materials shown in Fig. 2-20 produced a single lap shear specimen with a maximum strength of 40 MPa.

The main advantage of PCM welding over adhesive bonding methods is the speed at which joints can be assembled. A weld time of only 30 sec was necessary to achieve the maximum lap shear strength, and this could be reduced. Adhesive bonding techniques, on the other hand, take as long as several minutes. The bonding technology can be used with any material to which a thin layer of thermoplastic can be adhered prior to welding.

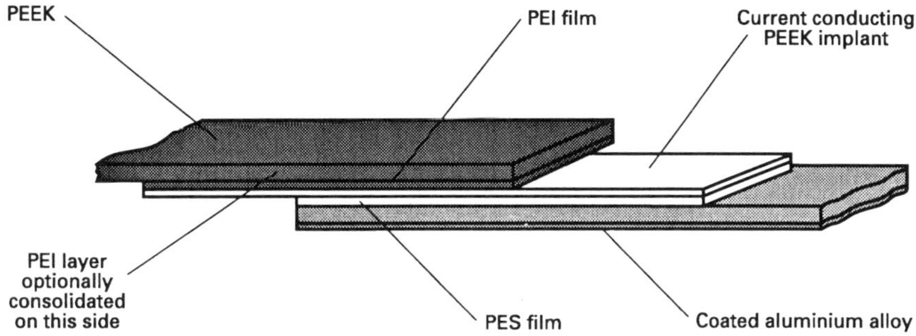

Fig. 2-20 In implant resistance welding, the bond can be strengthened by applying a polyether sulfone (PES) film to the aluminum surface and applying a polyether-imide (PEI) film to the polyether etherketone (PEEK). Courtesy of the Welding Institute

Industrial applications for components joined using PCM welding will have an upper operating-temperature limitation that stems from the glass transition temperature of the polymers involved. The Welding Institute recommends that a maximum temperature of 420 °C be used to achieve optimum results.

PCM welding has been used to join thermoplastic composite (APC-2) to alumina, L113-grade aluminum alloy to alumina, L113-grade aluminum alloy to AlN, and SiC-reinforced borosilicate glass composite to itself. The joints manufactured between APC-2 and L113-grade aluminum alloy were tested to a maximum lap shear strength of 30 MPa. For this lap shear strength, the joint required a peak current of 40 A, pressure of 2.67 MPa, and 30 sec (Ref 71).

Repair

Limited effort has gone into developing repair materials for composites, in contrast to the significant labor and expense that brought composites to the high position they hold today. This limited effort has been appropriate up until this point, but it is not appropriate for the future. The economic loss will be extremely great if we do not develop the repair materials of the future. A worse situation could be the loss of confidence in composites if they are not easily repaired. In particular, as composites gain wider acceptance on aircraft, there is a need to understand the requirements of the commercial airliners, general aviation, and the military regarding repair of these advanced materials.

In current practice, each composite part must be repaired using the original material specified by the manufacturer. For thermoset-based composites, each material has a limited shelf life (6 to 12 months) and must be stored in a freezer. Waste is high due to the difficulty of obtaining small amounts of material from manufacturers in a timely manner. Repair kits are available, but they are effective only if the part to be repaired is in a depot or the aircraft is on a maintenance cycle. The need to standardize repair materials is widely recognized (Ref 72-75). Material manufacturers, airline manufacturers, and commercial airliners are working to reduce the amount of duplication and increase the reliability of composite repairs (Ref 72).

References are available regarding the general use and selection of adhesives for specific uses, such as ship repair and the automotive industry (Ref 76-78). Also, previous studies have related the material properties of adhesives to joining performance (Ref 79-82). Armstrong (Ref 74), who did an extensive study relating adhesive properties to the lap shear performance of joints, expressed the need for standardized materials and more complete materials information from manufacturers. Papers on repair give a summary of materials that were used for specific applications or specific aircraft, but they generally do not address the issue of standardization (Ref 83-85).

Types of Damage. In the aircraft and helicopter industries, most repair relates to two types of damage: environmental factors (such as hail, lightning and bird strikes, and debris kicked up on takeoff or landing) and hanger rash (mishandling of aircraft or components on the ground). Puncture-type damage and microcracking of composites are common; damage of face skins can allow moisture ingression in the underlying honeycomb. Large-part or large-area repair can require replacement of the part or removal for remanufacturing. For damage over a smaller area, operators use prepregs and wet-lay-up techniques in bonded, bolted, or flush aerodynamic patching (Ref 86) (Fig. 2-21).

Yet another problem is that of personnel training and turnover. Evaluation of the type and extent of damage to a structure, and feasible approaches to repair given local material and equipment availability, demand an expertise that comes as much from hands-on experience as from formal training. This is compounded by the need to know the wide variety of composite structures already in use, each with its own design and lay-up schedule, which makes virtually every structure a special case (Ref 87).

Design Decisions. Several criteria determine how a damaged structure is to be repaired. The extent of the damage and the type of structure, primary or secondary, determine what is to be done. In most secondary structures, either a metal plate can be riveted or bolted to the composite or a two-part epoxy can be used with fabric reinforcement. These will be parts with minimal load requirements. Bolted repairs are the most common for battle-damage repairs because of their reliability. Alternative bonding methods require more rigorous control of surface treatments, storage of heat-sensitive bonding materials, and special equipment and/or expertise (Ref 87).

If a major structure is damaged, the airframe manufacturer or the services will generally want to bring the aircraft or part back to their facility to repair it. In most cases there will not be established procedures to follow, and engineering projects will have to be established to determine the validity of the repair. In most cases the original composite material is used in the repair. This can be done by pre-curing a patch and then bonding or fastening it in place. In other cases, if the part can be removed from the aircraft, the uncured patch can be built up and the part placed in an autoclave to cure. Most parts will have been made originally with a 177 °C curing system, which means that if the original material is used, a 177 °C cure will be required.

Though prone to many idiosyncrasies, field repair environments generally restrict users to vacuum-bag cure techniques (which yield low bonding pressures), localized heat sources such as heat blankets or pads, induction curing, infrared lamps, or hot air guns, and generally poor storage conditions. The materials to be used must be room-temperature storable for 1 year, curable at 177 °C or less within 2 h, and ideally be a one-component system. Typical repair information is shown in Table 2-3 and Ref 72 to 103.

Materials Selection. The key element for repairs is the selection of a suitable material combination that meets the demands of the composite material system used for production. For 125 °C and 175 °C curing of epoxy thermoset matrices, the Hysol EA 956 and 3M SP377 materials are very promising candidates. They feature room-temperature cure, the capability for low-, medium-, and high-temperature postcures, and low viscosity to allow for low-porosity laminates and resin injection.

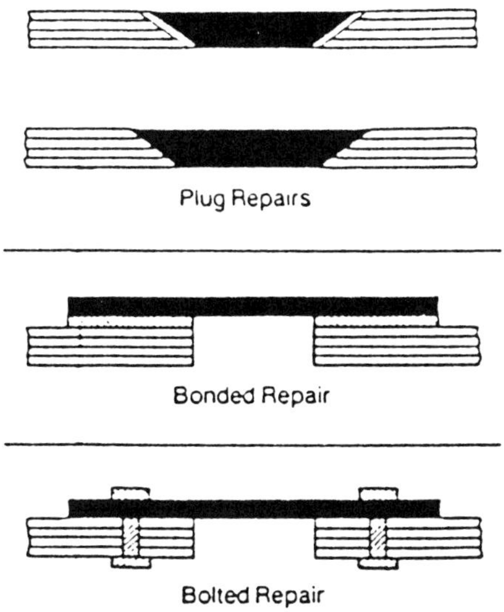

Fig. 2-21 Common methods of composite repair: placing a plug of material in the damaged area, or bonding or bolting on a patch

Table 2-3 Maximum repairable damage size for 737-300 aircraft

Primary structure		Secondary structure Repair method/cure temperature				
		Wet lay-up/postcure			Prepreg	
Item	Skin material	66 °C (150 °F)	93 °C (200 °F)	121 °C (250 °F)	121 °C (250 °F)	177 °C (350 °F)
Flap track fairings	Glass/epoxy 250 °F cured	5.0(a)	5.0(a)	...	No size limit	...
	Graphite/glass epoxy 350 °F cured	2.0 Interim only	(a)	5.0(a)(b)	...	No size limit
Wing tip Wing panels Aileron tab	Glass/epoxy 250 °F cured	5.0(a)	5.0(a)	...	No size limit	...
Aileron	Graphite/glass epoxy 350 °F cured	3.0 Interim only	...	5.0(a)	5.0(a)	No size limit
Spoiler	Graphite/epoxy 350 °F cured	3.0 Interim only	...	6.0(a)	6.0(a)	No size limit

Note: All values are in inches (1 in. = 25.4 mm). (a) 2 extra piles (per face sheet damaged) required on repair. (b) 150 °C (300 °F) only

Thermoplastics have been shown to be tougher and more damage resistant, but they have not been fully specified for repair applications in terms of adhesive bonding and reprocessing parameters. Economic tradeoffs will again come into play with the use of thermoplastics for repair, with the rapid, low-cost processability balanced against the significant capital already invested (but not yet amortized) for thermoset processing equipment such as autoclaves (Ref 86).

Process Selection. Xiao et al. (Ref 104) found that for repair of TPCs, fusion bonding offers significant advantages over conventional adhesive bonding and mechanical fastening. Induction heating is fast, efficient, and precise, and it has the potential for healing-type repairs as well as patching repairs. Resistance heating is equally fast and relatively simple. If current density, heating time, and consolidation pressure are controlled, resistance heating can produce lap shear strengths above 20 MPa in APC-2. Resistance heating is most adaptable for patching repair. Induction and resistance heating also allow control of the cooling rate, which affects the crystalline structure of a semicrystalline thermoplastic and, in turn, the mechanical properties of the composite.

Border et al. (Ref 105) also found that induction heating is fast. Portable power units were constructed that made induction heating well-suited for aircraft repair. The limitations on induction heating were its inability to heat a large area and a lack of control; however, research showed that a flexible coil could be constructed that would provide a striped heat zone. By bending the coil to fit the repair patch, the coil could be used on any type of repair without the need for skilled scanning of the patch. This coil represents a significant step forward in induction heating technology as complex curves and large areas can be repaired. By eliminating the need for scanning of the surface, control is also enhanced.

In studies of aircraft battle damage repair, Mahon et al. (Ref 106) found induction heating to be an excellent choice if combined with vacuum bag pressure as a repair technique.

A study by Davies et al. (Ref 107) examined two approaches to repair processing: fusion bonding and the application of conventional epoxy adhesives to TPCs. Carbon-PEEK composites were successfully assembled and repaired by both techniques, although both had some disadvantages. Adhesive bonding required an appropriate surface treatment, but the relatively low temperatures involved (180 °C) minimized the risk of damage to surrounding structure. Fusion bonding produced strong, tough joints, but locally high temperatures were necessary.

The use of other thermoplastic film adhesives enables fusion bonding to be applied at lower temperatures without the disadvantages of conventional adhesives, but careful selection of the thermoplastic is necessary if solvents are likely to be present.

Applications. Composite patch repairs on cracked metallic aircraft structures is an accepted technique for improving fatigue life and maintaining high structural efficiency. Boron-epoxy and graphite-epoxy composite patches, bonded with either room-temperature-cured or high-temperature-cured adhesives, attain sufficiently high fatigue lives to meet the damage tolerance requirements of many aircraft (Ref 108). In fact, the boron-epoxy patch with high-temperature adhesives gives better fatigue life than the graphite-epoxy patch with room-temperature adhesives. Several applications have been reported for this technique, such as Mirage 2I fighters, C-130 transporters, and F-111 fighters (Ref 109-113).

A recent first in-service composite material repair was applied to metal structure on the U.S. Air Force C-141B operational aircraft (Ref 114). Five locations were evaluated on the C-141B, representing areas susceptible to both fatigue and corrosion cracking. Repairs using either boron-epoxy or graphite-epoxy were beneficial for crack retardation, end fastener effect relief, and field stress reduction. The most obvious time-saving feature of these repairs was that no fastener removal or installation was involved. Extensive structural analysis using finite element modeling was done to ensure that the repair design would have the requisite strength and durability and to reduce the inspection burden after the aircraft was returned to service.

High-temperature PI composite materials are in use on military aircraft engines and are being considered for commercial aircraft engine applications. The titanium outer bypass duct on the GE 104 engine used on the F/A-18 aircraft has been replaced and is now manufactured as a duct from carbon/PMR-15*. In commercial applications, the first in-service prototype core cowl doors fabricated

* PMR-15 is a NADIC end-capped PI known as polymerization of monomer reactants (PMR). It was developed by NASA Lewis Research Center, Cleveland, OH, and is marketed by Dexter Hysol.

from PI composite materials are also being evaluated. Experience with carbon-epoxy composite parts used on the F/A-18 aircraft indicates that they are quite prone to damage induced by handling; however, they have proven to be very repairable. Parts manufactured from carbon/PMR-15 are expected to be damage prone as well. A program to develop the repair technology for this class of materials was jointly undertaken by Rohr Corp. and the Navy Air Development Engineering Installation at North Island, San Diego, CA (Ref 115). They found the following:

- Incorporation of steps at the patch edge, designed to reduce peak adhesive stresses, resulted in an increase in joint strength of 66% over the nonstepped version.
- A Rohr-modified PMR-15 resin used to bond precured laminates was not as strong as FM35 adhesive at 20 °C but was stronger at 260 °C.
- In general, laminates that were cocured without adhesive showed excellent shear strength at 20 and 260 °C, before and after thermal cycling.
- Thermal cycling for 1000 cycles between –17.8 and +232 °C (with a 10 min dwell at the extreme temperatures) had negligible effects on notched shear strength values. Long-term thermal cycling is expected to degrade properties.
- The best overall shear strength properties at 20 and 260 °C were obtained for Rohr-modified PMR-15. This resin has a very low methylene dianiline level (less than 0.1%), can be stored at 20 °C, and has mechanical and thermal properties that are equivalent or superior to those of standard PMR-15 (Ref 115).

Boron-epoxy composite materials have been fully qualified for primary structural applications on the F-14, F-15, and B-1 aircraft, as well as for patching and doubling on metallic structures to stop crack propagation and increase fatigue lives. The use of bonded boron-epoxy doublers for the repair of aircraft structures has considerable history in military aircraft. In fact, the first application of the technique was on the F-111 fighter/bomber wing pivot fitting, which dates to 1971. Since then there have been many more applications. In addition, the Air Force Materials Laboratory validated the concept and determined that adding bonded boron-epoxy doublers to cracked and damaged aluminum structures increases overall fatigue life a minimum of 20 to 25% (Ref 116). The use of bonded boron-epoxy composites, proven in the military, is a cost-effective way for commercial airframe manufacturers and airline maintenance facilities to extend the economic lives of their aircraft.

Boron-epoxy composite has been used to repair and structurally reinforce the forward titanium longeron of the B-1B bomber. The original remedy (drilling a hole at the end of each crack to inhibit propagation and bolting on an aluminum doubler) failed in 17 of 37 aircraft because drilling and bolting increases stress concentration, exacerbating the problem. The composite, applied in the form of prepreg tape and heat cured on the aircraft, was attached by adhesives that precluded drilling and reduced stress concentrations 15 to 20%. To reinforce the B-1B longeron, two doublers about 64 mm wide by 508 mm long were used. Repair time was less than two days, markedly less than by other methods and substantially reducing downtime.

Metal helicopter blades are repaired using carefully developed adhesive bonding techniques, so the procedure for repairing composite blades is similar. The primary stages are removal of damage, surface preparation, application of adhesive and repair material, and bonding under controlled cycle conditions. Conventional heating and vacuum or pressure bagging techniques are used in the repair of both metal and composite blades. In a similar manner to aircraft repair, metal blades may eventually be repaired using composite patching techniques. The best hope here appears to be boron, because this material is transparent to eddy current nondestructive testing and has high strength and modulus. The application to helicopters is currently in the demonstration and evaluation phase (Ref 117).

Future Repair Materials and Concepts. Currently no depot or field repair techniques exist for repair of high-temperature RMCs. High-temperature PI resins such as PMR-15 are used on structures whose operating temperatures are beyond the capability of epoxies and BMI. Existing repair materials cannot be applied to PMR-15 due to the higher service-temperature requirements, and existing repair procedures and methods cannot be applied due to the unique processing requirements of the material. The absence of repair materials would necessitate the removal, replacement, and scrappage of damaged parts.

In a series of evaluation tests performed by the Naval Air Development Center in Warminster, PA (Ref 118), a range of potential materials for high-temperature repair has been produced. The effort was directed at the development of patch materials that could be used for depot and field level repairs of 288 °C composite structures. The initial conclusion of this study was that processing concepts could be developed for the fabrication of preconsolidated formable patch materials for high-temperature composites. It was found that PMR-15 staged laminates could be produced, but that the material quality would be strongly dependent on the staging conditions used. The powder prepregged material form was determined to have the best laminate properties from the processes studied. The acetylenic PMR material showed promise as a vacuum-cure PI, but additional work will be required for the development of suitable staging conditions.

BMI-based composite materials are being applied extensively on structures requiring elevated-temperature service on emerging naval aircraft. Two repair approaches, one bolted and one bonded, were developed and demonstrated in a recent program. The bolted repair approach, which built on field-level maintenance experience with bolted repairs to metal structures, used a rapid-curing, elevated-service-temperature polythioether-based sealant. A full-scale bolted repair specimen was successfully structurally tested to validate this concept. The bonded repair approach employed patch materials that are storable at room temperature and can be processed with the support equipment currently in the field. Fully consolidated, partially reacted, vitrified patch materials were fabricated through a combination of advanced staging and nonautoclave processing procedures. Several conclusions can be drawn from the study (Ref 119):

- The nonautoclave/staging process was successfully applied to a number of BMI composite materials, and the laminates produced by this technique were comparable in microstructure and mechanical properties to those produced with autoclave processing techniques.
- The nonautoclave/staging procedure provided a technique for the fabrication of ambient-temperature-storable, high-quality patches for repair of BMI composite components.
- Evaluation of the two repaired specimens showed that the patch approach restored full structural integrity to BMI laminate materials. The bonded repair restored 170% of design ultimate strain to the damaged specimen. The bolted repair restored 0.005 in./in. strain, which extended the design ultimate strain requirement of 0.004 in./in.

There is a trend in the aerospace community toward an increased use of elevated-temperature composites in the leading edges of supersonic aircraft and in structures surrounding the engine bay. These structures have complex shapes and highly curved surfaces that are difficult to repair using staged prepreg patches or bolted-on metal plates. In order to repair complex structures, a wet lay-up patch is required, according to Mehrkam and Cochran (Ref 120). This patch will provide the capability to fabricate a highly conformable repair. BMIs are typical resins used in composite structures exposed to temperatures of 204 °C in wet environments. However, BMIs are difficult to use in a wet lay-up procedure because they are solid at ambient temperature. The BMI resin must be heated so that it will become a low-viscosity liquid capable of impregnating dry fabric.

Recently, cyanate esters have shown promise as wet lay-up resins for use at high temperatures. A dicyanate ester monomer was investigated by Mehrkam and Cochran for the repair of high-temperature composite structures (Ref 120). The resin chosen was a low-viscosity liquid that could easily impregnate a fiber cloth without the addition of heat. The reaction chemistry, thermal stability, and mechanical properties of the dicyanate ester system were studied and evaluated in this work. The study also evaluated the nonautoclave processing characteristics of AroCy L-10 from Rhône Poulenc.

The Ideal Repair Material. It is appropriate to establish what the repair material of the future should be, and then compare some of the newer materials to the ideal. The repair material of the future should have a significant number of characteristics, including, but surely not limited to, the following:

- It should be a one-part prepreg, so that repair technicians don't have to mix up and impregnate the repair material with a two-part epoxy, for the same reasons that one-part prepregs are used almost exclusively for the original parts rather than wet systems.

- It should be in a form that makes it easy to do the repair. This could mean preplied patches ready to apply, with a specified lay-up and orientation, and including adhesive layers, or a variety of other forms.
- It should have extremely good, broad, room-temperature storage time. In many cases repairs will need to be made where refrigeration does not exist.
- It should cure at as low a temperature as possible. The ideal cure would minimize distortion of the existing structure, would not tend to have internal pressure buildup from moisture, and would be a low-pressure cure due to part removal and size.
- It should replace the properties that exist in the original composite material. This is the most important characteristic of a future repair material. The ideal condition would be for the repair material to replace the original material ply for ply, with the same number and orientation. Otherwise the transition or interface between the undamaged material and the repair must have sufficient shear strength to transfer the required loads. (One method of doing this is a scarf joint.)

The idea of a generic repair material must also be addressed. Is it truly mandatory to use the same material that was used in the original structure? Aren't the characteristics of the material the important factors? The development of a material that could be used on all aircraft and all parts, no matter what the original material, is vital to the future of repaired PMCs.

REFERENCES

1. M. Bowtell, Advances Presented in Plastics and Composites Joining, *Adhes. Age*, Aug 1991, p 42-44
2. *Adhes. Age*, May 1993, p 40-41
3. N.E. Rouse, Improved Methods for Thermoplastic Bonding, *Mach. Des.*, 11 April 1985, p 72-79
4. W.S. Johnson, Ed., *Adhesively Bonded Joints: Testing, Analysis, and Design*, ASTM, 1988
5. A.V. Pocius and R.P. Wenz, *SAMPE J.*, Vol 21 (No. 5), 1985, p 50
6. T. Stevens, Joining Advanced Thermoplastic Composites, *Mech. Eng.*, March 1990, p 41-45
7. J. Osterndorf, R. Rosty, and M.J. Bodnar, Adhesive Bond Strength and Durability Studies Using Three Different Engineering Plastics and Various Surface Preparations, *SAMPE J.*, Vol 25 (No. 4), July-Aug 1989, p 15-19
8. P.W. Rose and E.M. Liston, Treating Plastic Surfaces with Cold Gas Plasmas, *Plast. Eng.*, Oct 1985, p 41-45

9. "Adhesive Bonding of Vectra Liquid Crystal Polymers," Letter RCH-36-49, Celanese Corp., 7 Oct 1986
10. "Adhesive Bonding of Torlon Parts," Bull. TAT 36A, Amoco Chem. Co., Chicago, IL
11. K. Rogus and J. Aragon, "Plasma Etch: A More Controlled Surface Preparation for Bonding Composite Structures," 34th Int. SAMPE Symp. & Exhib. (Reno, NV), 8-11 May 1989
12. E. Occhiello, M. Morra, G.L. Guerrini, et al., Adhesion Properties of Plasma-Treated Carbon/PEEK Composites, *Composites,* Vol 23 (No. 3), May 1992, p 193-200
13. T.J. Kinloch and B.R.K. Blackman, "Continuing Challenges Facing Adhesive Technologist," Symposium of Society of Chemical Industry and British Institute of Metals (London), March 1993.
14. "Tech Monitoring, Materials Technologies, Structural Adhesives," SRI International, 1991, p 95
15. A.B. Carter, III, Preserving the Structural Integrity of Advanced Composite Materials through the Use of Surface Mounted Fasteners, *SAMPE J.,* Vol 25 (No. 4), July-Aug 1989, p 21-25
16. *Advanced Composites Design Guide,* 3rd ed., Air Force Materials Laboratory, 1973
17. Joining of Advanced Composites, DARCOM-P 706316, *Engineering Design Handbook,* U.S. Army Material Development and Readiness Command, March 1979
18. W.A. Lees, A Review—The Design and Assembly of Bonded Composites, Chapter 36, *Composite Structures,* I.H. Marshall, Ed., 6th Int. Conf., Elsevier Applied Science, 1991, p 471-506
19. E.E. House, "Integrally Stiffened Laminate Construction," 9th Nat. SAMPE Tech. Conf. (Atlanta, GA), 4-6 Oct 1977
20. "Composite Rear Fuselage Manuf. and Methods & Technology Program," Tech. Prop. SPB79-A2664-RD, Sikorsky Aircraft, Div. of United Tech., 21 Aug 1979
21. *DOD/NASA Structural Composites Fabrication Guide,* Vol II, 2nd ed., Contr. F33615-77-C-5256, Lockheed Co. Manuf. Technology Div., Air Force Materials Laboratory, May 1979
22. "F-15 Composite Wing Flight Test," AFML Contr. F33615-71-6-1576, AFML-TR-75-78, McDonnell Douglas
23. E.B. Birchfield and R. Kollmansberger, "Develop Fabrication/Processing Techniques for High Temperature Advanced Composites for Use in Aircraft Structures," AFML-TR-72-91, AFML Contr. F33615-70-C-1546, McDonnell Douglas, July 1972
24. R.E. Politi, Recent Developments in Polyimide and Bismaleimide Adhesives, *High Temperature Polymer Matrix Composites,* T.T. Serafini, Ed., Noyes Data, Park Ridge, NJ, 1987, p 123-127
25. N. Allbee, Adhesives for Structural Applications, *Adv. Compos.,* Nov-Dec 1989, p 42-50
26. D.R. Bradshaw, Structural Adhesive Q&A: Performance, Cost and Use, *Adhes. Age,* Sept 1989, p 44-48

27. M. Bowtell, Adhesive Activities in United Kingdom Include Seminars, *Adhes. Age*, Sept 1989, p 67-69
28. J.D. Buckley, R.L. Fox, and J.R. Tyeryar, Seam Bonding of Graphite Reinforced Composite Panel Using Induction Heating, *Ind. Heat.*, March 1987, p 32-34
29. Joining Composite Structures in Space, *Aerosp. Eng.*, July 1989, p 9-11
30. A.S. Brown, Thermoplastic Composites—Material of the '90s?, *Aerosp. Am.*, Jan 1990, p 28-33
31. M.M. Schwartz, *Composites Materials Handbook*, 2nd ed., McGraw-Hill, 1992
32. T.J. Reinhart, Structural Bonding Needs of Aerospace Vehicles, *Adhes. Age*, Jan 1989, p 26-28
33. W.H. Schwartz, Fasteners Hold their Own, *Assembly Eng.*, April 1989, p 24, 25
34. Fasteners for Composites, *Aircr. Eng.*, July 1988, p 2, 3
35. L.H. McCarty, Lightweight Fasteners Reduce Unit Stresses on Composites, *Des. News*, 4 Sept 1989, p 186, 187
36. Manufacturers are Developing Improved Structural Systems, *Aviat. Week Sp. Technol.*, 12 Dec 1988, p 67-72
37. M. Dubble, Composite Fasteners Face the Aerospace Challenge, *Mach. Des.*, Vol 61 (No. 14), 6 July 1989, p 102-104
38. F.E. Smode, *Composite Material Fastener Reduces Lightning Dangers*, D. Whittaker, Ed., Sterling Publications, 1992, p 63-65
39. Hi-Lex, a trademark of Hi-Shear Corp.
40. R.Y. Kim, Fatigue Strength, *Engineered Materials Handbook*, Vol 1, *Composites*, ASM International, 1987, p 436-444
41. J. Rausch, Fastening Fundamentals for Fabricated Composites, *Assembly Eng.*, Oct 1988, p 34-36
42. N. Allbee, Aspects of Mechanically Fastening Advanced Composites, *Adv. Compos.*, Nov-Dec 1988, p 54-60
43. Composite Fasteners, *Rotor and Wing*, May 1993, p 53
44. H. Gumbleton, Hot Gas Welding of Thermoplastics—An Introduction, *Joining Mater.*, Vol 5, 1989, p 215-218
45. B.E. Turner and J.R. Atkinson, Repairability of Plastic Automobile Bumpers by Hot Gas Welding, *ANTEC '89: 47th Ann. Tech. Conf.*, 1-4 May 1989
46. R.J. Wise, Design of a Magnetostrictive Ultrasonic Transducer, *TWI J.*, Vol 32 (No. 4), 1991, p 81-84
47. A.J. Cheney and W.E. Ebeling, Methods for Joining Plastic Parts, 14th Ann. Tech. Conf., Society of Plastics Engineers, Jan 1958
48. H. Potente and H. Kaiser, Process Variant of Vibration Welding with Variable Welding Pressure, *Advances in Joining New Structural Material*, Proc. Internat. Conf. of the International Institute of Welding (Montreal), 23-25 July 1990
49. M.N. Watson, Hot Gas Welding of Thermoplastics, *Weld. Inst. Bull.*, May-June 1989, p 77-80

50. Hot-Gas Welding System Consolidates Complex Thermoplastics Composites, *Adv. Mater.*, Vol 11 (No. 16), 25 Sept 1989, p 4
51. M. Girardi, Plastics at TWI—An Update, Bulletin 5, *TWI J.*, Sept/Oct 1992, p 100-103
52. R.A. Grim, Fusion Welding Techniques for Plastics, *Weld. J.*, March 1990, p 23-28
53. J. Powers and W. Trzaskos, "Recent Developments in Adhesives for Bonding Advanced Thermoplastic Composites," 34th Int. SAMPE Symp. & Exhib. (Reno, NV), 8-11 May 1989
54. D.M. Maguire, Joining Thermoplastic Composites, *SAMPE J.*, Vol 25 (No. 1), Jan-Feb 1989, p 11-14
55. V.K. Stokes, Vibration Welding of Thermoplastics, Parts, I, II, III, and IV, *Polym. Eng. Sci.*, Vol 28 (No. 11), 1988
56. G. Kempe, H. Krauss, and G. Korger-Roth, Adhesion and Welding of Continuous Carbon-Fiber Reinforced Polyether Etherketone (CF-PEEK/APC-2), *Developments in Science and Technology of Composite Materials,* Proc. Fourth European Conf. on Composite Materials (Stuttgart), 25-28 Sept 1990, European Association for Composite Materials, Elsevier, 1990, p 105-112
57. N.S. Taylor, Ultrasonic Welding Plastics, *Weld. Inst. Bull.*, July-Aug 1988, p 159-162
58. W.H. Schwarz, Ultrasonic Sound Off, *Assembly Eng.*, Jan 1990, p 24-27
59. A. Benatar and T.G. Gutowski, "Ultrasonic Welding of Advanced Thermoplastic Composites," Ind. Liaison Prog. Rep. 6-41-88, Massachusetts Institute of Technology, March 1988
60. J. Wolcott, Recent Advances in Ultrasonic Technology, *ANTEC '89: 47th Ann. Tech. Conf.*, Society of Plastics Engineers, 1-4 May 1989, p 502-505
61. R.J. Wise, Approaching Total Joining Technology for Carbon Fibre Reinforced Thermoplastics, Bulletin 1, *TWI J.*, Jan/Feb 1992, p 4-6
62. A. Leatherman, Induction Bonding Finds a Niche in an Evolving Plastics Industry, *Plast. Eng.*, Vol 4, 1981, p 27-29
63. A.J. Smiley, "A New Concept for Fusion Bonding Thermoplastic Composites," Joining of Composites Conf., (Garden Grove, CA), 28-29 March 1989, Society of Manufacturing Engineers
64. E.M. Silverman and P.A. Griese, Joining Methods for Graphite/PEEK Thermoplastic Composites, *SAMPE J.*, Vol 25 (No. 5), Sept-Oct 1989, p 34-38
65. M.J. Day and M.F. Gittos, The Application of Microscopy to Welded Joints in Thermoplastics, *TWI J.*, 390/1989, Jan 1989
66. R.C. Don et al., Fusion Bonding of Thermoplastic Composites by Resistance Heating, *SAMPE J.*, Jan-Feb 1990, p 59-66
67. E.C. Eveno and J.W. Gillespie, Jr., "Resistance Welding of Graphite Poly Ether Ether Ketone Composites; An Experimental Investigation," CCM88-35, University of Delaware, Dec 1988
68. E.C. Eveno, "Experimental Investigation of Resistance and Ultrasonic Welding of Graphite Reinforced Poly Ether Ether Ketone Composites," CCM88-30, University of Delaware, Dec 1988

69. W.J. Cantwell, P. Davies, and P.E. Bourban, et al., Thermal Joining of Carbon Fibre Reinforced PEEK Laminates, *Compos. Struct.*, Vol 16, 1990, p 305-321
70. Connect, *TWI J.*, May 1992, p 3
71. R.J. Wise, Polymer Coated Material (PCM) Welding: A Novel Technique for Joining Dissimilar Materials—Preliminary Study, Bulletin 1, *TWI J.*, 454/1992, July 1992
72. "Guidance Materials for the Design, Maintenance, Inspection and Repair of Thermosetting Epoxy Matrix Composite Aircraft Structures," International Air Transport Association, 1991
73. J.A. Renbert and J.C. Seferis, Fundamentals of Composite Repair, EM92-100, *SME Conf. Comp. in Mfg. '92* (Anaheim, CA), 7-8 Jan 1992
74. K.A. Armstrong, personal communication, July 1989
75. J. Thorbeck, *Advanced Materials: Cost Effectiveness, Quality, Control, Health and Environment*, SAMPE/Elsevier Science, 1991
76. R. Mohan, *Plast. Eng.*, Feb 1990
77. R.C. Allen, J. Bird, and J.D. Clarke, *Mater. Sci. Technol.*, Vol 4, Oct 1988
78. W.A. Lees, *Mater. Sci. Technol.*, Vol 4, Dec 1988
79. H.N. Cheng, *Adhes. Age*, Dec 1988
80. N.S. Taylor, Welds in Thermoplastic Composite Materials, *Developments in Science and Technology of Composite Materials*, Proc. Fourth European Conf. on Composite Materials (Stuttgart), 25-28 Sept 1990, European Association for Composite Materials, Elsevier, 1990, p 325-332
81. D. Bamborough and P.M. Dunckley, *Adhes. Age*, Nov 1990
82. C.O. Arah, D.K. McNamara, H.M. Hand, et al., *SAMPE J.*, Vol 25 (No. 4), July/Aug 1990
83 F. Lee, S. Brinkerhoff, and S. McKinney, 20th Int. SAMPE Tech. Conf., 27-29 Sept 1988
84. R. Raghava, *J. Polym. Sci. B, Polym. Phys.*, Vol 26, 1988
85. T.J. Reinhart, WL-TR-91-4054, Third DOD/NASA Composites Repair Technology Workshop, May 1991, 14-17 Jan 1991
86. V.P. McConnell, In Need of Repair, *Adv. Compos.*, May-June 1989, p 60-70
87. L.K. English, Field Repair of Composite Structures, *Mech. Eng.*, Sept 1988, p 37-39
88. L.C. Cook, The Repair of Aircraft Structures Using Adhesive Bonding—A Survey of Current UK Practices, *TWI J.*, Vol 392, 1989
89. R. Hall, M.D. Raizenne, and D.L. Simpson, Proposed Composite Repair Methodology for Primary Structure, *Composites*, Vol 20 (No. 5), Sept 1989, p 479-483
90. C. Ramsey, "Battle Damage Repair of Composite Structures," 34th Int. SAMPE Symp. and Exhib. (Reno, NV), 8-11 May 1989
91. T. Reinhart, "Composites Supportability in the U.S. Air Force," 34th Int. SAMPE Symp. and Exhib. (Reno, NV), 8-11 May 1989

92. M. Chichon, "Repair Adhesives: Development Criteria for Field Level Conditions," 34th Int. SAMPE Symp. and Exhib. (Reno, NV), 8-11 May 1989
93. T. Steelman, "Repair Technology for Thermoplastic Aircraft Structures (REPTAS)," 34th Int. SAMPE Symp. and Exhib. (Reno, NV), 8-11 May 1989
94. B. Purcell, "Battlefield Repair of Bonded Honeycomb Panels," 34th Int. SAMPE Symp. and Exhib. (Reno, NV), 8-11 May 1989
95. C. Hartup, "Battle Damage Repair of Thermoplastic Structure," 34th Int. SAMPE Symp. and Exhib. (Reno, NV), 8-11 May 1989
96. G. Sivy and P. Briggs, "Rapid Low-Temperature Repair System for Field Repair," 34th Int. SAMPE Symp. and Exhib. (Reno, NV), 8-11 May 1989
97. J. Connolly and D. Vannice, "Development of PMR-15 Repair Concepts," 34th Int. SAMPE Symp. and Exhib. (Reno, NV), 8-11 May 1989
98. M. Livesay, N. Smith, and E. Castenada, "Fast Simple System for Field Repair of Composite Armor and Structures," 34th Int. SAMPE Symp. and Exhib. (Reno, NV), 8-11 May 1989
99. M. DiBerardino, J. Dominquez, and R. Cochran, "Bonded Field Repair Concepts Using Ambient Temperature Storable Materials," 34th Int. SAMPE Symp. and Exhib. (Reno, NV), 8-11 May 1989
100. R. Meade, "Repair of Thermoplastic Composites," 34th Int. SAMPE Symp. and Exhib. (Reno, NV), 8-11 May 1989
101. D. Bittaker, "Manufacturing Technology for Bonded Repair Procedures," 34th Int. SAMPE Symp. and Exhib. (Reno, NV), 8-11 May 1989
102. J. Mahon, "Induction Bonded Repair of Aircraft and Structure," 34th Int. SAMPE Symp. and Exhib. (Reno, NV), 8-11 May 1989
103. M.S. Sennett, "Field Repair of Composite Materials in Army Service; Planning for Future," TR-89-45, FR5/89, Materials Test Lab, Watertown, MA, May 1989
104. X.R. Xiao and S.V. Hoa, Repair of Thermoplastic Composite Structures by Fusion Bonding, *35th Int. SAMPE Symp.* Vol 35, Society for the Advancement of Material and Process Engineering, 1990, p 37-43
105. J. Border, R. Salas, and M. Black, Induction Heating Development for Aircraft Repair, *35th Int. SAMPE Symp.*, Vol 35, Society for the Advancement of Material and Process Engineering, 1990, p 1411-1418; WL-TR-91-4054, *Third DOD/NASA Composites Repair Technology Workshop*, May 1991, p 82-90
106. J. Mahon, S. Stracquadaini, and M. DiBerardino, Rapid Processible Materials for Aircraft Battle Damage Repair (ABDR), WL-TR-91-4054, *Third DOD/NASA Composites Repair Technology Workshop* (Wright-Patterson Air Force Base, Dayton, OH), May 1991, p 617-625
107. P. Davies, W.J. Cantwell, P.-Y. Jar, et al., Joining and Repair of a Carbon Fibre-Reinforced Thermoplastic, *Composites*, Vol 22 (No. 6), Nov 1991, p 425-431
108. C.-L. Ong and S.B. Shen, The Reinforcing Effect of Composite Patch Repairs on Metallic Aircraft, *Int. J. Adhes. Adhes.*, Vol 12 (No. 1), Jan 1992, p 19-25

109. A.A. Baker, Repair of Cracked or Defective Metallic Aircraft Components with Advanced Fiber Composites—An Overview of Australian Work, *Compos. Struct.*, Vol 6, 1984, p 153-181
110. A.A. Baker, R.J. Calinan, R.J. Davis, et al., Repair of Mirage III Aircraft Using the BFRP Crack-Patching Technique, *Theor. Appl. Fract. Mech.*, Vol 2, 1984, p 1-15
111. F.A. Sandow and R.K. Cannon, "Composite Repair of Cracked Aluminum Alloy Aircraft Structure," Report AFWAL-T87-3072, Air Force Wright Laboratories, 1987
112. A.A. Baker and R. Jones, *Bonded Repair Aircraft Structures*, Martinus Nijhoff, Dordrecht, Netherlands, 1988
113. L. Molent, L. Callinan, and R. Jones, Design of an All Boron/Epoxy Doubler Reinforcement for the F-111C Wing Pivot Fitting Structural Aspects, *Compos. Struct.*, Vol 11, 1989, p 57-83
114. T.F. Christian Jr. and D.O. Hammond, Composite Material Repairs to Metallic Airframe Components, *J. Aircraft*, Vol 29 (No. 3), May-June 1992, p 470-476
115. R. Ghaffarian and D. Perl, Development of Repair Procedures for High Temperature Carbon/Polyimide Composite Materials, WL-TR-91-4054, *Third DOD/NASA Composites Repair Technology Workshop* (Wright-Patterson Air Force Base, Dayton, OH), May 1991, p 678-690
116. R.P. Caruso, *Boron/Epoxy Composites for Metallic Aircraft*, D. Whittaker, Ed., Sterling Publications, 1992, p 48-50
117. G. Marsh, Repairing Rotor Blades, *Aerosp. Compos. Mater.*, Vol 4 (No. 2), May 1992, p 26-29
118. M.F. DiBerardino, P.A. Mehrkam, and T.M. Donnellan, Repair Concepts for Polyimide Composite Structures, NADC, Structural Composites, *Proc. 6th Ann. ASM/ESD, Advanced Composites Conf.* (Detroit), 8-11 Oct 1990, ASM International, p 659-666
119. M.F. DiBerardino, E.L. Rosenzweig, and T.M. Donnellan, Repair Concepts for Bismaleimide Composite Structures, NADC, Structural Composites, *Proc. 6th Ann. ASM/ESD, Advanced Composites Conf.* (Detroit), 8-11 Oct 1990, p 641-650
120. P. Mehrkam and R. Cochran, Liquid Dicyanate Ester Monomer Resin for Elevated Temperature Composite Repair Applications, NADC, *Proc. Am. Soc. for Composites 7th Tech. Conf.* (University Park, PA), 13-15 Oct 1992, p 12-20

BIBLIOGRAPHY

- "Adhesive Bonding Composite Materials," PB93-864163/WMS, Nerac, Inc., April 1993
- *Adhesive Bonding Handbook for Advanced Structural Materials*, ESA PSS 03-210 Issue 1, European Space Research & Technology Centre, Netherlands, Feb 1990

- R. Allred, "Effect of Plasma Treatment of Carbon/BMI Interfacial Adhesion," 34th Int. SAMPE Symp. & Exhib. (Reno, NV), 8-11 May 1989
- L. Bastien, I. Howie, R.C. Don, et al.,"Manufacture and Performance of Resistance Welded Graphite Reinforced Thermoplastic Composite Structural Elements," CCM 90-33, University of Delaware, Oct 1990
- A. Benatar, R.V. Eswaran, and S.K. Nayar, "Ultrasonic Welding of Thermoplastics, Part I: Near-Field," MR8902, Edison Welding Institute, Columbus, OH, Feb 1989
- A. Benatar and T.G. Gutowski, A Review of Methods for Fusion Bonding Thermoplastic Composites, *SAMPE J.*, Jan-Feb 1987, p 33-39
- A. Benatar and Z. Cheng, "Ultrasonic Welding of Thermoplastics, Part II: Far-Field," MR8903, Edison Welding Institute, Columbus, OH, April 1989
- M. Benfeldt, "Vibration Welding of Thermoplastics," Session 1, ASM Int. Ann. Mtg. (Cincinnati, OH), 21-24 Oct 1991
- Bismaleimide Resins: High-Performance Thermosets Await Boost from Military Aircraft, *High Technol. Mater. Alert*, Vol 6 (No. 9), 5-6 Sept 1989
- B.R. Bonazza, Adhesive Bonding of Improved Polyphenylene Sulfide Thermoplastic Composites, *35th Int. SAMPE Symp.*, 2-5 April 1990, p 859-865
- J.H. Brahney, Fasteners for Composite Structures Examined, *Aerosp. Eng.*, June 1985, p 8-14
- C.-C. Chen, Mechanical Fasteners for Thick Laminates, *Proc. 4th Ann. Conf. on Advanced Composites* (Dearborn, MI), ASM International, 13-15 Sept 1988, p 391-400
- D.J. De Renzo, *Advanced Composite Materials: Products and Manufacturers*, Noyes Data, 1988
- J.G. Dillard and I. Spinu, Plasma Treatment of Composites for Adhesive Bonding, *Proc. 4th Ann. Conf. on Advanced Composites* (Dearborn, MI), ASM International, 13-15 Sept 1988, p 199-208
- R.C. Don, J.W. Gillespie, Jr., and C.L.T. Lambing, Experimental Characterization of Processing-Performance Relationships of Resistance Welded Graphite/Poly Ether Ether Ketone Composite Joints, *Polym. Eng. Sci.*, Vol 32 (No. 9), 1992, p 620-630
- R.C. Don, et al., "Fusion Bonding of Thermoplastic Composites by Resistance Heating," CCM89-19, University of Delaware, Aug 1989
- B.K. Fink, R.L. McCullough, and J.W. Gillespie, Jr., "Heating of Continuous Carbon Fiber Thermoplastic Composites by Magnetic Induction," WL-TR-91-4054, Third DOD/NASA Composites Repair Technology Workshop, May 1991, p 64-72
- Focused Infrared Bonds, Shapes TP's, *Plast. Technol.*, Jan 1989, p 32-41
- B. Grimm, "Thermoplastic Welding Processes," Session 1, ASM Int. Ann. Mtg. (Cincinnati, OH), 21-24 Oct 1991

- S.T. Holmes, R.C. Don, J.W. Gillespie, Jr., et al., "Sequential Resistance Welding of Large-Scale IM7/PEEK Double Lap Joints," CCM 91-55, University of Delaware, Dec 1991
- S.T. Holmes and J.W. Gillespie, Jr., Thermal Analysis for Resistance Welding of Large-Scale Thermoplastic Composite Joints, *Proc. 7th Tech. Conf.* (University Park, PA), 13-15 Oct 1992, American Society for Composites, p 135-146
- W.D. Humphrey and S.T. Peters, "Joining Methods for Filament Wound Structures," Session 5D, 38th Int. SAMPE Symp. and Exhib. (Anaheim, CA), 10-13 May 1993
- R.L. Keller and W.R. Mull, "Supportability Improvements in Advanced Composites Field Repair," WL-TR-91-4054, Third DOD/NASA Composites Repair Technology Workshop, May 1991, p 13-25
- C.L. Lambing, "Advancement in Welding Technology for Polymer to Metallic Joints," Session 5D, 38th Int. SAMPE Symp. and Exhib. (Anaheim, CA), 10-13 May 1993
- A.M. Maffezzolli, J.M. Kenny, and L. Nicolais, Welding of PEEK/Carbon Fiber Composite Laminates, *SAMPE J.*, Vol 25 (No. 1), Jan-Feb 1989, p 35-39
- J. Magenson, "Recent Advances in Ultrasonic Plastic Welding Joint Design and Process Control," Session 1, ASM Int. Ann. Mtg. (Cincinnati, OH), 21-24 Oct 1991
- J.M. Marinelli and C.T. Lambing, "A Study of Surface Treatments for Adhesive Bonding of Composite Materials," Session 4C, 38th Int. SAMPE Symp. and Exhib. (Anaheim, CA), 10-13 May 1993
- L. McCarthy, Part Design is Critical for Good Welds, *Plast. World*, June 1990, p 61-67
- S.H. McKnight, S.T. Holmes, and I. Howie, "Resistance Heated Dual Resin Bonding," CCM-92-43, University of Delaware, 1992
- L.D. Michelove, R.F. Caruso, and P. Adams, Advanced Materials: Looking Ahead to 21st Century, *22nd Int. SAMPE Tech. Conf.*, Vol 22, 1990, L.D. Michelove, R.F. Caruso, and P. Adams, Ed., Society for the Advancement of Material and Process Engineering
- M.G. Murch, Hot Plate Welding Injection Molded Thermoplastics: The Effect of Process Parameters, *TWI J.*, 389/1989, Jan 1989
- C.-L. Ong, Y.-C. Wu, and C.-L. Lee, "Repair of Graphite/Epoxy Horizontal Stabilizer," Session 4F, 38th Int. SAMPE Symp. and Exhib. (Anaheim, CA), 10-13 May 1993
- R.B. Ostrom, S.B. Koch, and D.L. Wirz-Safranek, Thermoplastic Composite Fighter Forward Fuselage, *SAMPE Q.*, Vol 21 (No. 1), Oct 1989, p 39-45
- P.F. Packman, H.J. Hietala, and K.C. Shultz, "Fastener and Hole Imperfection Effects on the Bolted Joint Strength of Graphite/Epoxy Laminates,"

Session 1B, 38th Int. SAMPE Symp. and Exhib. (Anaheim, CA), 10-13 May 1993

- R.J. Pilarski and M. Chookazian, Ford Bonds Thermoplastics with Electromagnetic Welding, *Adhes. Age*, June 1986, p 20, 21
- "Polyimide Adhesives," PB93-863215/WMS, Nerac, Inc., April 1993
- Robotic Infrared Heating Technique Used for Joining PEEK Composites, *Reinf. Plast.*, Vol 33 (No. 2), Feb 1989, p 48
- N.E. Rouse, Improved Methods for Thermoplastic Bonding, *Mach. Des.*, 11 April 1985, p 72-79
- D. Ruffner and R. Patel, "Evaluation of Low Temperature Curing Repair Adhesives," Session 4F, 38th Int. SAMPE Symp. and Exhib. (Anaheim, CA), 10-13 May 1993
- M. Sasdelli, V.M. Karbhari, and J.W. Gillespie, Jr., The Design and Use of Molded-in Metal Inserts and Attachments in Resin Transfer Molding, *Proc. 8th Advanced Composites Conf.* (Chicago), ASM International, 2-5 Nov 1992, p 193-198
- R.R. Schmitt and W.J. Horn, Viscoelastic Relaxation in Bolted Thermoplastic Composite Joints, *35th Int. SAMPE Symp.*, 2-5 April 1990, p 1336-1347
- T. Schneider, Reversible Bonding of Composite Aircraft Structures, SAWE-2055, *51st Ann. Conf.*, Society of Allied Weight Engineers, May 1992
- B.A. Stein, et al., "Rapid Adhesive Bonding Concepts," TM 86256, NASA Langley Research Center, June 1984
- B.A. Stein, et al., "Thermoplastics/Thermosetting Adhesive Specimen Bonding," Case-LAR-13066-1, National Aeronautics and Space Administration, 29 May 1985
- H.D. Swartz and J.L. Swartz, "Focused Infrared Melt Fusion: Another Option for Welding Thermoplastic Composites," EM89-175, Joining Composites Conf. (Garden Grove, CA), 28-29 March 1989, Society of Manufacturing Engineers
- N.S. Taylor, "A Study of the Ultrasonic Welding of Short Glass Fibre Reinforced Polyphenylene Sulphide," Members Rep. 388/1989, Welding Institute, Jan 1989
- Thermoplastics Inch Ahead, *Plast. Technol.*, March 1992, p 44-45
- S.M. Todd, Joining Thermoplastic Composites, *22nd Int. SAMPE Tech. Conf.*, 6-8 Nov 1990, L.D. Michelove, R.F. Caruso, and R. Adams, Ed., Society for the Advancement of Material and Process Engineering, p 383-392
- J.A. Vaccari, Processing Metal-Matrix Composites, *Am. Mach.*, March 1985, p 95-97
- S.S. Volkov, "Welding of Plastics," FTD ID(RS)T 0104-88, Air Force System Command, April 1988

- R. Walsh, M. Vedula, and M.J. Koczak, Comparative Assessment of Bolted Joints in a Graphite Reinforced Thermoset vs. Thermoplastic, *SAMPE Q.*, July 1989, p 15-19
- S.-Y. Wu, Adhesive Bonding of Thermoplastic Composites, *35th Int. SAMPE Symp.*, 2-5 April 1990, p 846-858
- X.R. Xiao, S.V. Hoa, and K.N. Street, Processing and Resistance Welding of APC-2 Composite, *J. Compos. Mater.*, Vol 26 (No. 7), 1992, p 1031-1049
- R. Yosomiya, K. Morimoto, A. Nakajima, et al., *Adhesion and Bonding in Composites*, Marcel Dekker, 1990
- N.K. Young, C.-C. Chen, and W.L. Cheng, Joining Sections of Thick Fiberglass Laminate, *Proc. 4th Ann. Conf. on Advanced Composites* (Dearborn, MI), ASM International, 13-15 Sept 1988, p 209-216

Chapter 3

Joining MMCs, CMCs, CCCs, and Specialty Composites

Recent years have seen significant developments in new high-strength, lightweight materials. Although these materials were originally developed for the aerospace and defense industries, development and improvement has continued and some are now being considered for high-volume markets, particularly automotive components and civil aircraft.

The most widely developed MMCs consist of a light alloy matrix containing a high-strength ceramic reinforcement. Particles, whiskers, or continuous fibers can be used for reinforcement, and the volume fraction of the reinforcement is typically between 10 and 60%. These materials usually have high strength and high moduli, and they retain their strength to high temperatures, so that operations such as forging and machining are difficult. As a result, joining simple parts is an attractive method for making complex products.

There are a number of problems peculiar to MMCs:

- In the case of particulate reinforcement, the particles may have different densities from the matrix, and this can lead to pronounced segregation effects.
- Undesirable chemical reactions can take place between the particle and the matrix, especially when the matrix is molten. This has been shown to be a particular problem in aluminum alloys reinforced with SiC, and it can lead to formation of undesirable aluminum carbides. This problem is illustrated in the micrograph in Fig. 3-1, which shows the joint line of a resistance weld between a 6082 aluminum alloy (6082 Al) and 6082 Al reinforced with SiC particles. Segregation of the particulates can be clearly seen, but no evidence was found in this case of reactions between the matrix and SiC particles. Correct manipulation of the welding parameters can minimize segregation.
- The ductility of the materials is very low compared with that of the matrix material, and thus the risk of cracking in the weldment, particularly in fusion processes, is higher.

Fig. 3-1 Resistance weld between 6082 Al alloy and 6082 Al alloy reinforced with SiC. 25×

These problems can be overcome by careful optimization of the joining process and by applying a detailed understanding of the metallurgical phenomena peculiar to these materials.

Joining Processes

With all new materials, a need to join them to the same or other materials is inevitable, and it can be argued that the industrial potential of the materials will be severely limited if joining procedures cannot be developed. Ideally, joining technology should provide a variety of processes suitable for manufacturing high-integrity joints with optimum mechanical properties and ease of inspection. It is unlikely that all these criteria can be fully satisfied for MMCs in the short term.

Solid-State Processes

Several solid-state processes have been investigated, evaluated, and tested.

Friction welding (FRW) has proved very successful in making sound joints. Two components are moved against each other under pressure, developing frictional heat. The relative motions can be of several types (e.g., rotary, orbital, or linear). The Welding Institute (Ref 1) has examined the rotary and linear motions; however, the rotary motions have been most widely used. A soft layer is formed at the interface, and this undergoes intense plastic deformation, much of the material being extruded from the joint. Once this layer has formed, a solid-state bond exists between the two components, and it is normal to allow the bond to cool under pressure.

From a metallurgical point of view, examination of the interface (which is often hard to find exactly) shows a continuous microstructure with no evidence of microstructural discontinuities across the interface. Sometimes there is evidence of fragmentation of the reinforcement, caused by repeated impacts during the friction process, but this phenomenon is not always observed (Fig. 3-2).

So far, only limited mechanical property data have been obtained (Ref 1), but the results appear encouraging. Yield values in cross-weld tensiles of up to 90% of parent material value have been obtained. It is probable that suitable postweld heat treatment would give even better results, because the age-hardening matrix would have been subjected to solution treatment or overaging, depending on location and thermal cycle.

Other investigations have found that the continuous-drive FRW process is a useful method for joining of SiC_p/Al-MMCs. Table 3-1 shows the yield and tensile strength, ductility, and Vickers hardness in the heat-affected zone (HAZ) of friction welded composites. Corresponding data for the base materials (B) are included for comparison. Unlike conventional welding methods, FRW of aluminum-base composites produces no pores or clusters of SiC particles in the HAZ. In general, the plastic deformation occurring during the welding operation has no significant effect on the overall distribution of SiC particles in the material. This, in turn, implies that the stiffness (E-modulus) is maintained across the joint. It therefore follows that the resulting HAZ strength level is mainly controlled by reactions taking place within the aluminum matrix during the weld thermal cycle. As a result, full recovery of HAZ strength can be achieved by the use of an appropriate postweld heat treatment, which involves solution heat treatment at 535 °C followed by artificial aging at 160 °C for 10 h.

Inertia friction welding (IFRW), somewhat similar to friction welding, has been investigated in several studies as a joining media for SiC/6061 Al and Al_2O_3/6061 Al. The former material was studied for possible incorporation in the

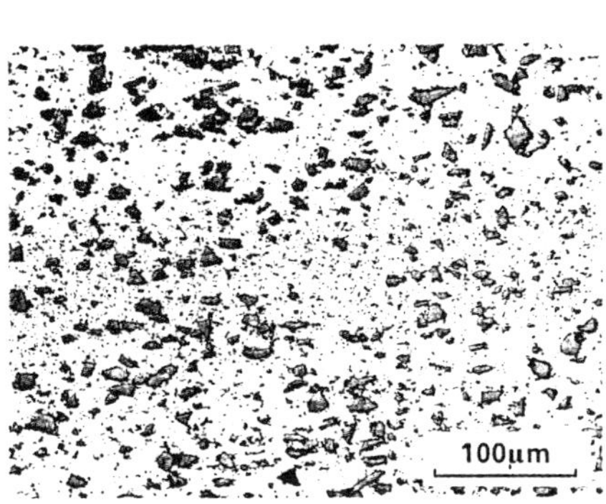

Fig. 3-2 Section through the bond line of a friction weld in 6061 Al/SiC/20_p

Table 3-1 Mechanical properties of extruded composite base metal (B) and friction welded composite metal (a to h) for two different temper conditions, T3 and T6

Temper	Tensile strength, MPa	Yield strength, MPa	Elongation, %	HV_5
T3/(B)	109.9	212.2	11.4	60
T3/(B′)	111.4	211.9	10.6	60
T3/(a)	100.8	177.2	15.9	60
T3/(b)	100.8	179.0	14.5	60
T3/(c)	313.8	331.8	1.7	135
T3/(d)	308.1	347.0	2.6	135
T6/(B)	315.1	351.9	3.7	135
T6/(B′)	315.1	351.2	3.5	135
T6/(e)	207.7	265.6	2.7	90
T6/(f)	206.6	270.2	3.2	90
T6/(g)	311.1	346.7	3.1	135
T6/(h)	314.6	348.4	...	135

structural components of lightweight torpedoes. The MMC under consideration (23 wt% SiC) offers a very favorable strength-to-weight ratio, has near-isotropic properties, and is very heat resistant (about 65% more tensile strength than unreinforced 6061 Al at 260 °C).

One naval study has involved the pistons of the MK 46 torpedo engine, which are made of 6061 Al alloy. Operation of the engine subjects the piston tips to intense heating and causes melting and erosion in the upper portion. Current practice is to replace the pistons after five torpedo runs. To reduce maintenance and associated costs resulting from the relatively short operational life of these pistons, the feasibility of replacing the top portion of the piston with a SiC/Al-MMC, using inertia welding as the joining method, was investigated. Inertia welding was selected because it is a low-cost procedure that permits the joining of dissimilar metals and requires no special environment. The cylindrical geometry of the pieces to be mated allows use of this technique with very little preparation and no special tooling.

The IFRW process has been widely used for joining dissimilar materials and exhibits several characteristics that make it a viable candidate for welding aluminum MMCs and dissimilar materials to conventional monolithic aluminum alloys (Ref 2). These characteristics include an absence of melting at the weld surface, rapid heating and cooling rates, and the expulsion of severely heat-and-deformation affected material during the final forging and upset stage of the process. Mechanical mixing and diffusion occurs between thin layers of metal on each side of the weld interface during IFRW of monolithic materials. Under ideal conditions, this local mixing and deformation is associated with a stable plastic condition across the entire interface region. Principal welding parameters (rotational velocity, inertial mass, and axial force) must be carefully controlled to take full advantage of these process characteristics.

A major problem with fusion welding of Al-Al_2O_3 materials is particulate coalescence and a nonuniform particulate distribution in the fusion zone. Although somewhat controversial, it is generally accepted that melting does not occur at the interface during IFRW (Ref 2). In addition, the production of high-joint-efficiency welds in alumina-particulate-reinforced composites requires retention of the particulates and grain structure across the weld interface. This microstructure retention is promoted in IFRW by the extremely high heating and cooling rates. These high heating rates and short periods at peak temperature tend to reduce or eliminate particulate-matrix reactions as well as promote a uniform particulate distribution and reduce grain coarsening near the weld interface.

In one investigation (Ref 2), IFRW was explored as an alternative to fusion welding of particulate-reinforced aluminum MMCs because of its solid-state nature. The purpose was to identify the procedural variables that influence the weldability of 6061 Al reinforced with Al_2O_3 particulates during IFRW. Extruded rods of 6061 Al/Al_2O_3/15_p underwent IFRW. The aluminum MMC material was joined in the extruded-and-annealed condition (64:1 ratio) as well as in the fully heat-treatable condition, T6 (Fig. 3-3). A commercially available heat of 6061-T651 extruded rod was used for the dissimilar welding tasks and as a control alloy throughout the test program. Joint properties were evaluated with a series of tensile and reduced-radius bend tests. The microstructures in the bond region were characterized in detail by using conventional light microscopy. Fractographic analysis of the fracture surfaces subsequent to mechanical testing was also performed with scanning electron microscopy (SEM). The findings were as follows:

- Tensile tests conducted on similar and dissimilar materials revealed joint properties that were approximately 40 to 50% of those of the parent materials, respectively (Fig. 3-4).
- Reduced-radius bond tests showed limited ductility of the similar and dissimilar MMC joints.
- Metallographic analysis of the weld region showed good interfacial mixing at the bondline (Fig. 3-5). Alumina particulates were continuous along the bond interface. The bond region appeared to be sound and free of any defects.
- Fractographic examination of fracture surfaces from tensile specimens revealed distinct regions of ductile tearing between the Al_2O_3 particulates. Discrete areas of nonbonding were also observed on the fracture surface. They coincided with the outward radial flow of material from the joint interface and are believed to result from particulate matrix interaction during upsetting.
- The process variables, such as the welding pressure and the speed required to achieve desirable amounts of flash and near-base-metal proper-

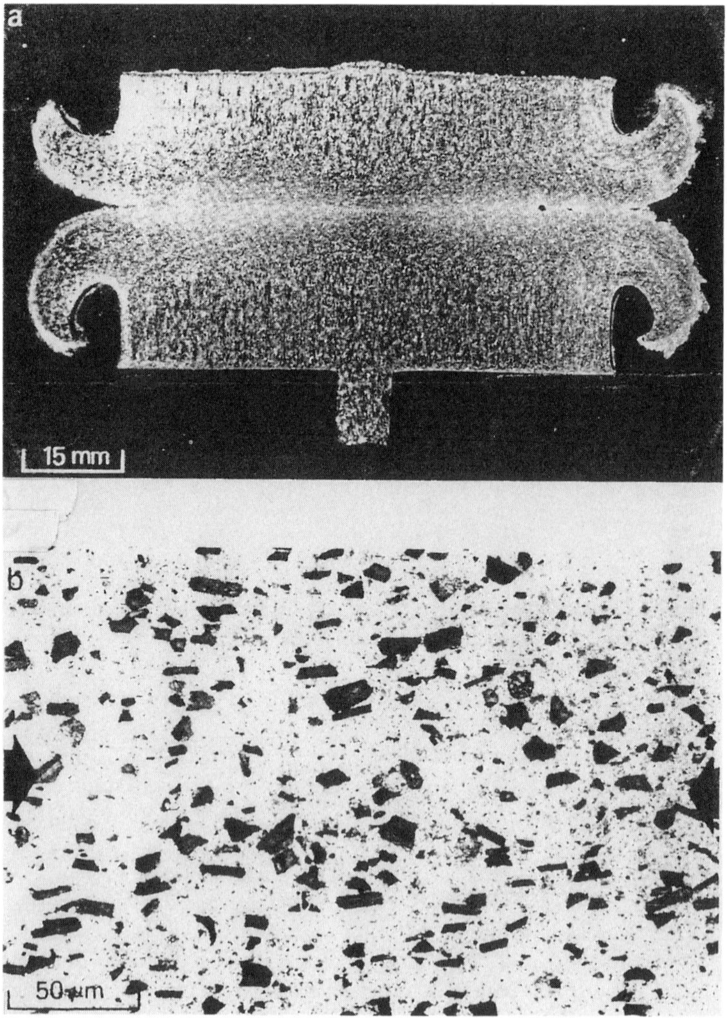

Fig. 3-3 Weld of Al/Al_2O_3-T6 to Al/Al_2O_3-T6. (a) Macrograph of weld joint. (b) Micrograph of weld interface microstructure. Note good alloy mixing along weld interface.

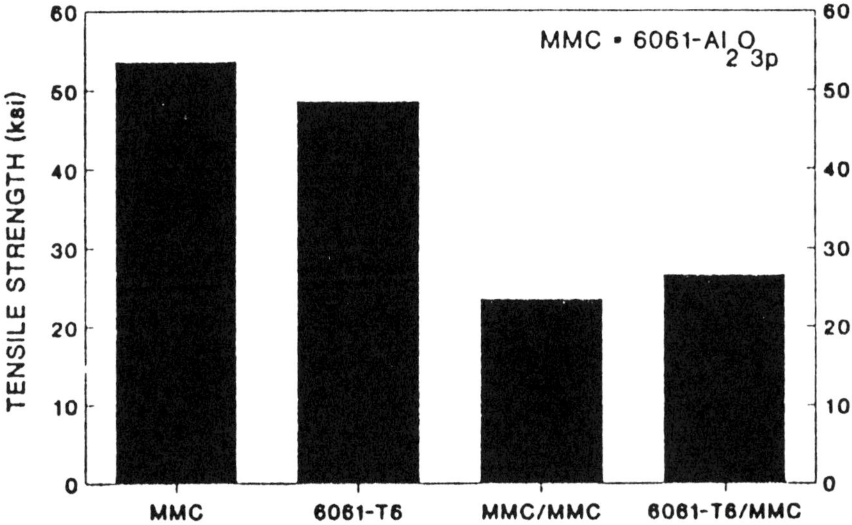

Fig. 3-4 Tensile test results for similar and dissimilar material combinations. 10 ksi = 70 MPa. Source: Ref 2

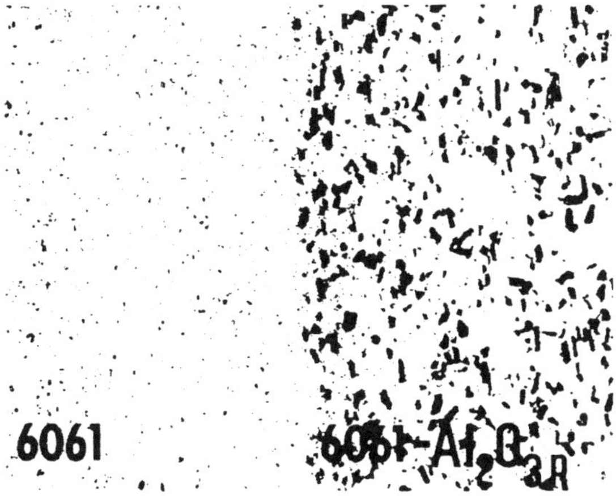

Fig. 3-5 Micrographs of the bond region in 6061 Al/6061 Al-Al_2O_3 material. Source: Ref 2

ties, were lower than those needed for joining unreinforced aluminum alloy 6061-T6.

- The welds between composite and monolithic aluminum exhibited room-temperature strengths of 83% of those of the composite base material.
- Light microscopy revealed that fracture locations were predominantly associated with the region near the weld interface in the composite material. Hardness surveys corroborated that fracture occurred in regions of low hardness near the weld interface.
- SEM analysis of failed tensile specimens confirmed the presence of a varying fracture topography.
- The tensile strength of the welds exhibited high joint efficiencies when high axial pressure and flywheel speed were used.

Diffusion Welding. At the Welding Institute, preliminary trials of diffusion welding on pure Al/SiC/20p and 6061 Al/SiC/20p produced successful bonds using copper, silver, or no interlayers. Metallographic examination confirmed the need for precise control of the process variables, to ensure that no particulate-rich zones or particulate-free zones form at the interface.

Mechanical data on the process is limited, but it is known that a bond that appears sound when examined microstructurally does not necessarily give good mechanical properties. Diffusion bonding of aluminum is not easy, because of the tenacious and stable oxide, so it is difficult to achieve reproducibly satisfactory bonds in the matrix alloys. Addition of the reinforcing particulate imposes further restrictions on process variables.

A recently completed development program documented the ability of low-temperature, solid-state bonding (LTSSB) to produce high-quality joints between MMC parts. The researchers found that the composite materials could be bonded at temperatures and pressures low enough to prevent dimensional change (deformation) and/or damage of the composite materials. Thus, net-shape bonding could be accomplished without the need for subsequent machining. The bonding could be performed at locations, such as space, where subsequent machining would be impractical or impossible. Room-temperature shear strengths of the resultant bonds ranged from about 13.8 to over 1380 MPa. Except for the case where aluminum or magnesium alloys were bonded to themselves, the LTSSB joints were always stronger than the shear strengths of the composites themselves.

A joint technology known as fiber-reinforced advanced titanium was recently announced. The hybrid system consists of the selective incorporation of a carbon core substrate (SiC fibers or other similar high-strength, high-modulus fibers) into the face sheets of a three-sheet superplastic forming/diffusion bonding (SPF/DB) sandwich structure of Ti-6Al-4V. With fiber volume fractions of 7 to 40%, local increases of 30 to 40% in strength and 100% in selective stiffness have been demonstrated (Fig. 3-6). Other metal-matrix materials that have been investigated include commercially pure titanium, Ti-3Al-2.5Sn, and Ti-6Al-2Sn-4Zr-2Mo. The fibers being evaluated besides SiC include B_4C-coated boron and Borsic.

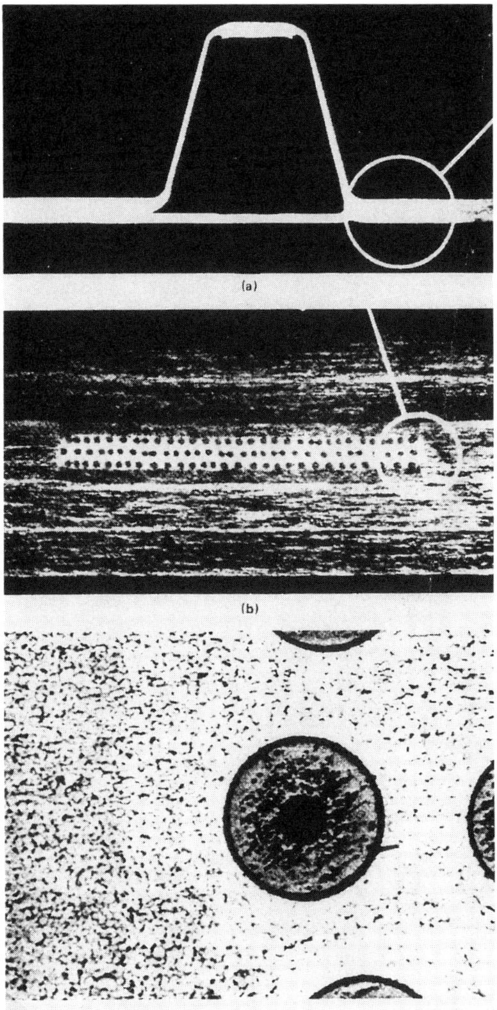

Fig. 3-6 Details of selectively reinforced titanium structures. (a) End view of sectioned bar. (b) Placement between sheets. Material is B_4C-coated boron in a Ti-6Al-4V superplastic forming/diffusion bonding hat-stiffened section. 10×. (c) Diffusion bond. 200×. Source: Ref 13, 14

The major finding thus far is that considerable degradation in strength can be expected after consolidation of MMC and thermal exposure (899 °C for several hours) during SPF/DB. However, no significant degradation in a modulus has been observed. The SiC laminates show a dramatic increase in compressive modulus over the matrix alloy, whereas B_4C-coated boron shows the highest potential for achieving both strength and stiffness after processing. Consolidation of prefabricated tapes into structures has led engineers to consider diffusion bonding (as an integral part of the SPF/DB process) and brazing of the composite laminates after SPF/DB panel fabrication. The SPF/DB process gives the structural designer additional design freedom at no additional cost. It permits joining of two or more sheets in numerous ways. The addition of the low-strain fibers greatly reduces the engineering elongation and toughness of the MMC compared to that of the unreinforced metallic material.

Fusion Processes

Fusion processes are economical and flexible, and they generally require only modest capital investment. They are well established for aluminum alloys and are attractive for joining MMCs. However, experience has highlighted a number of problems that must be considered.

The welding of Al-MMCs differs from the welding of monolithic aluminum alloys in that the weld pool of the composite contains a solid phase, because the reinforcement does not necessarily melt at welding temperatures. Thus, in the strictest sense, the composite weld pool is actually a partially melted zone. The most obvious effect of the presence of the reinforcement is the increased viscosity of the Al-MMC weld pool (Ref 3). This weld pool does not flow and wet as readily as that of monolithic aluminum alloys. Mass transfer is limited because of the viscous nature of the MMC weld pool.

Therefore, heat flow by convection in the weld pool is believed to be less effective than in aluminum alloys, and conductive heat flow through the aluminum alloy matrix is thought to play a larger than normal role in determining the temperature distributions and cooling rates in and around the weld pool of Al-MMCs. The differences in heat flow in the MMC, relative to that in monolithic aluminum alloys, can affect the resulting microstructures and the stress distributions in the MMC weld. The use of other reinforcements, such as Al_2O_3, which is stable in molten aluminum, will overcome the problem.

A second problem is rejection of the particulate by the solidifying interface. Most MMCs are reasonably homogeneous, but following fusion welding, ceramic particles agglomerate in the interdendritic regions, reducing their efficiency as a reinforcement. The magnitude of this effect and its significance are clearly influenced by the particle size and the cooling rate, which determines the dendritic cell size. In large weld pools, gravitational effects may also lead to enhanced macrosegregation, although there have been no known investigations of this.

With certain particle-matrix combinations, there is a significant risk of particle-matrix interactions. In particular, SiC will react with aluminum at high temperatures to produce an aluminum carbide, Al_4C_3. This phase is extremely brittle, but it is also unstable and reacts with any moisture to form acetylene. Experience has shown that the life expectancy of welds where this reaction has occurred can be very short, perhaps only days. However, the reaction can be suppressed by using high-silicon alloy matrices or by avoiding high energy density processes that raise the peak weld pool temperature.

A third problem may occur with some MMCs made by powder metallurgy techniques. If the occluded gas content is too high, gas will come out of solution at high temperatures, causing extensive cracking in the HAZ. However, this does not appear to be a problem with current powder route MMCs (Ref 1).

Gas-Metal Arc Welding and Gas-Tungsten Arc Welding. At the Welding Institute, GTAW in an 8090 Al/SiC/20p alloy, using a 4047A filler metal, resulted in extreme lack of mixing. However, Duralcan has demonstrated that MMCs can be joined with GTAW or GMAW with procedures chosen to minimize dilution, thus overcoming the poor miscibility of the molten MMC.

Alcan International is in the prototype stage of welding 101.6 mm diameter composite driveshafts to aluminum yokes at speeds of 152.4 cm/min. Individual welds are being completed in 11 to 12 s. The composite consists of 6061 $Al/Al_2O_3/20_p$. The composite material is easier to weld than neat or unreinforced aluminum because the Al_2O_3 makes the weld pool more viscous. The seamless tubes, with wall thickness of 2 mm, were manufactured from extrusion billets 270 mm in diameter. The stiffness of the 6061 $Al/Al_2O_3/20_p$, which is the primary reason for using this material, is not influenced by heat treatment and is only a function of the volume of Al_2O_3 in the composite. Hence, a wide range of strengths, ductilities, and stiffnesses can be produced by varying the mechanical processing, the heat treatment, and the volume fraction of reinforcement. To fabricate the aluminum composite tubes into assemblies, the forged 6061 Al yokes were welded to the ends of the tubes. The GMAW process with 5356 Al filler wire was used in the standard driveshaft assembly fixtures to produce components for evaluation.

MSNW Inc. (Ref 4) conducted an exploratory fusion welding program under government sponsorship that examined welding procedures for 6061 Al/Al_2O_{3p} composite base plates using primarily the GMAW process. Continuous composite welding electrodes, 1.14 mm in diameter, were successfully produced by extrusion/rolling/drawing. Welds were prepared using short-circuit transfer and spray transfer GMAW procedures employing conventional 4043 Al and composite 4043 $Al/Al_2O_3/10_p$ welding electrodes. The welding performance of both the conventional and composite 4043 Al welding electrodes on the composite base plate was unsatisfactory. There was little retention of Al_2O_3 in either the conventional or composite electrode welds, and there was extensive weld metal porosity and cracking in the composite electrode welds made by either the short-circuit transfer or spray transfer welding process. This was held to be largely due to poor wettability of the base plate and the dispersed Al_2O_{3p} by the 4043 Al filler metal.

It was then determined that magnesium-containing filler metal provided significant improvements in both weld metal soundness and Al_2O_{3p} retention. The as-welded mechanical properties of welds produced with a conventional magnesium-containing 5356 Al electrode were found to be superior to those achieved with the conventional 4043 Al electrode. Magnesium additions to welds produced with the 4043 Al/Al_2O_{3p} composite electrode improved both soundness and Al_2O_3 retention. Magnesium-containing filler metal compositions such as 5356 Al appear to be promising starting points for the construction of composite electrodes for GMAW welding of Al-Al_2O_{3p} composite base plates and form the basis of continued research. Other compositions that might be suitable on the basis of their wetting characteristics toward Al_2O_3 particulates are compositions that contain copper and silicon, such as A390 (Ref 4).

The University of Wisconsin-Milwaukee used manual GTAW and aluminum filler metal to carry out some preliminary work on the weldability of a 6061 Al/SiC/15_p MMC (Ref 5).

Sporting goods, as well as automobiles and aerospace applications, are promising MMC markets. An example is the Specialized Bicycles M-2 mountain-bike frame, which uses welded tubes made from Duralcan 6061 Al/$Al_2O_{3p.}$ A fork-brace component is another MMC part. MMC proponents contend that the ease with which these materials are cast and forged lends both the materials and the process to component production more readily than conventional fiber-reinforced composites. Raleigh has an SP-2000 road bike with three of the main tubes made of 6061 Al/SiC_p.

Resistance Welding. There is a phenomenon that so far defies rational explanation: a mass segregation effect found in some resistance welds, in which the SiC particulates segregate preferentially to the periphery of the molten nugget (see Fig. 3-1). It is not clear whether this is caused by a hydrodynamic effect generating a centrifugal force or whether it is a surface tension effect. At any rate, the explanation is of academic interest only, as the effect can be suppressed by careful control of welding parameters (Ref 1).

Both resistance welding and forge welding are quite adaptable to a range of discontinuously reinforced aluminum composites. Problems associated with welding these materials typically have two causes. For the SiC-reinforced Al-MMCs, the high temperatures and relatively long times at temperature permit dissolution of the carbides and subsequent precipitation of detrimental phases (Al_4C_3). For the Al_2O_3-reinforced Al-MMCs, similar conditions allow the individual reinforcing particles to agglomerate, which reduces material performance.

Al-MMCs are generally spot welded using conditions similar to those for conventional aluminum alloys. However, these materials typically require substantially less current (about half) for welding. This is probably due to the increase in bulk resistivity associated with the addition of the particles.

It is believed that the minimal reaction of the SiC to form Al_3C_3 is due to two factors. First, in resistance welds, the temperature of the liquid nugget only barely exceeds the melting point of the material (Ref 6), providing little, if any, driving force for the reaction. Second, temperature excursions above the liquidus are relatively short, typically less than 0.1 sec, which minimizes any potential reaction.

Spot welds on discontinuously reinforced aluminum composites do show some redistribution of the reinforcement in the fusion zone. Typically, the fusion zone is relatively free of the reinforcement near the fusion line, with an increased density of the reinforcement near the center of the weld. It is possible that this redistribution of the composite is the result of macroscopic liquid flow during solidification, but the effect is still under study. The effect of this reinforcement-free zone on the mechanical performance of the weld is not known.

In the fuselage section designed for the national aerospace plane, there are hundreds of "hat-type" structural shapes of SiC_f/Ti that are spot welded to 1.2 m^2 composite panels. The MMC is expected to withstand temperatures up to 760 °C and is considered critical to the primary fuselage structure of the national aerospace plane.

Laser Beam Welding (LBW). Pulsed laser welding of A356 Al/SiC/10_p and A356 Al/SiC/20_p composite was used by University of Tennessee Space Institute scientists (Ref 7) to produce diverse microstructures in and around the weld. Nonequilibrium and unconventional phases such as Al_4C_3, $Al_4Si_2C_5$, and several unidentified phases evolved during the processing. The surface of the SiC particles was physically and chemically modified. The extent of microstructural changes was at minimum in the low-duty-cycle welds (50% and 57%) and at maximum in the high-duty-cycle welds (80% and 91%). Control of the heat input, and thereby control of the microstructure in and around the joint region, was achieved by operating the laser in the pulsed mode. The composite processed with intermediate-duty cycles represented the optimum changes in microstructure and possessed optimum strengths. Tensile tests of the longitudinal welds showed enhancement of mechanical properties (tensile strength and percent elongation), whereas tests of transverse welds showed degradation of the strength and ductility of both composites.

Kawall and Viegelahn (Ref 8) investigated the feasibility of welding an Al_2O_3-reinforced 6061 Al alloy. They found that in joining this composite material by conventional welding and using a filler metal, the joint was left with a lesser number of Al_2O_3 particles. Therefore, laser welding was tried as an alternative. The results were that:

- Al_2O_3-reinforced 6061 Al alloy coupled better than the plain 6061 Al alloy at low power. However, above 3.5 kW, coupling deteriorated and a considerable portion of the beam was lost to reflection.

- Plasma suppression was a major problem in composite laser welding. Gas flow over the weld had to be very closely controlled to achieve the plasma suppression. For best results, the gas flow must be turbulence-free and the welding direction should be such that the newly formed weld moves with the flow of the shield gas.
- Loss of reinforcing Al_2O_3 particles in the weld zone was another problem in the laser welds. This loss of particles was more pronounced in the keyhole region and near the top of the weld zone.
- Mechanically sound welds were not obtained in this preliminary study. However, welds obtained without the use of filler materials showed further promise in laser welding of the composite material.

Electron beam welding (EBW) appears to have potential for joining at least some SiC-reinforced Al-MMCs. Like LBW, the EBW process is capable of producing rapid thermal cycles. However, the physics of beam-material interaction differs for the two processes. Heating during LBW results from the absorption of photons by the substrate, whereas heating during EBW occurs by the transfer of kinetic energy to the atoms of the substrate via collisions with the high-energy electrons of the beam.

A follow-up study to the previously discussed LBW work (Ref 9) determined that EBW can be used successfully to join the same cast discontinuously reinforced aluminum (A356 Al/SiC/15_p) without the formation of large amounts of Al_4C_3 (Ref 10). Electron beam welds and CO_2 laser beam welds were produced at identical energy inputs, travel speeds, and focused beam diameters. The use of EBW resulted in sound welds with a uniform distribution of reinforcement and a fine grain size. The electron beam welds exhibited an order of magnitude less Al_4C_3 than the laser beam welds made using the same parameters. The electron beam welds made at 85 mm/sec with sharp beam focus contained almost no Al_4C_3. Electron beam welds made at slower speeds and/or with defocused beams contained somewhat more Al_4C_3.

The different mechanisms of energy transfer in EBW and LBW appear to affect the final weld microstructures of SiC-reinforced Al-MMCs and, thus, the weld properties. This occurs despite the fact that both welding processes allow for rapid thermal cycles with low overall heat input. Unfortunately, EBW cannot be successfully applied to all SiC-reinforced Al-MMCs. For example, EBW on a wrought 2014 Al/SiC/15_p MMC cut the test part. It is not known whether the problem resulted from the manufacturing process (powder metallurgy) or was inherent to the material.

Capacitor-discharge (C-D) resistance spot welding has much potential for the joining of aluminum, magnesium, and titanium MMCs. This process uses the electrical energy discharged from a bank of capacitors to rapidly heat the weld zone via I^2R heating*. Compared with conventional resistance welding processes, C-D resistance welding is characterized by an extremely short weld thermal cycle, typically less than 10 to 20 msec in duration. This short cycle promotes the concentration of the weld in a small zone near the weld interface, rather than allowing it to be conducted away from this region into the surrounding base material. The process is also characterized by an extremely rapid cooling rate, which further minimizes heat effect to the surrounding material. The C-D welding process has been successfully applied to the following aluminum and magnesium MMCs:

- 6061 Al/SiC/ up to 40_p
- 6061 Al/$B_4C/30_p$
- AZ61 Mg/$B_4C/40_p$
- 6061 Al/SiC/48_f
- 2024 Al/$B_4C/30_p$

The above weld microstructures exhibit no porosity and no observable alteration of the SiC and B_4C reinforcement.

A recently completed study (Ref 11) showed that solid-state and fusion welds have been produced between sheets of monolithic and SiC-fiber-reinforced Ti-6Al-4V using C-D resistance spot welding. Several conclusions were developed:

- Capacitor voltage (and corresponding current) and electrode force could be controlled to reproducibly generate solid-state and fusion spot welds in both monolithic and fiber-reinforced Ti-6Al-4V sheet.
- Solid-state and fusion welds in Ti-6Al-4V sheet exhibited tensile shear strengths above the minimum requirements set forth in military specification MIL-W6858D. Shear fracture in the solid-state welds occurred through the weld interface region, whereas fracture in the fusion welds occurred by nugget pullout.
- Some of the solid-state welds produced in the fiber-reinforced sheet exhibited complete welding across the weld interface with no evidence of fiber displacement or degradation. Increased voltage promoted the initiation of melting at the fiber-matrix interface and, at sufficiently high voltage, across the entire weld interface. Excessive interface melting promoted appreciable fiber dissolution and displacement.
- Optimized solid-state welds in the fiber-reinforced material exhibited tensile strength and shear stress levels that were 60 and 80%, respectively, of similar welds produced in the monolithic material. This reduction in

* It should be noted that during the resistance spot welding of titanium, I^2R heating (I, current; R, resistance) occurs principally within the titanium sheets due to the high resistivity of titanium and not at the faying surfaces due to a high interface resistance. This is in contrast to steel and aluminum, for which heating at the faying surfaces is much more important.

fracture strength was attributed to fracture along a layer of fibers in the weld HAZ adjacent and parallel to but remote from the weld interface.

- Dissimilar solid-state and fusion welds were produced between monolithic and fiber-reinforced sheets. The average tensile shear strength and stress of the welds produced at 160 V were 70 and 80%, respectively, of similar welds produced in the monolithic material. Fracture of all welds occurred remote from the weld interface along an adjacent layer of fibers.

Work on the C-D welding of Al-MMCs has also been reported (Ref 12). The capacitive system used in this work was basically a forge-butt welding process. Specifically, the system used oriented parts in a butt configuration, with the ends separated by a projection on one of the surfaces. A force was then applied to the workpieces and the energy from the discharging capacitors was introduced. The result was surface melting, followed by impact and forging of the components. This work has shown that Al_4C_3 formation can be precluded on several types of SiC-reinforced Al-MMCs. Consistent with the other resistance welding processes, the short exposure times and rapid cooling rates (approximately 10^6 °C/sec) were responsible for the retention of the SiC.

Cast welding is a new method that has been developed for metallurgically bonding inserts into a cast component during a normal casting operation. Recently, there has been interest in making pistons or other cast components that are strengthened in critical areas by ceramic reinforcement (e.g., SiC particulates or Al_2O_3 short fibers). This method consists of making a preform of the ceramic reinforcement, placing it in the die, and squeeze casting or applying high pressures in order to infiltrate the ceramic preforms and form a composite component. This is often unsatisfactory because of preform deformation, positioning, component size limitations, and expense.

The only other currently used method for producing a component made from two different materials is to fabricate the cast and composite portions separately and perform a subsequent welding or brazing operation. However, it is very difficult to weld MMCs, and the long times involved in brazing can cause severe degradation of the properties of both the composite and the cast structure. Also, it may be impossible to access the joint location in order to perform the weld.

Using cast welding, successful bonds have been obtained:

Between cast Al-12wt%Si and inserts of:

Al-12wt%Si/SiC/55_p
Al-12wt%Si/SiC/15_p
2014 Al/Al_2O_3/15_p
6061 Al/Al_2O_3/15_p
A356 Al/SiC/15_p
ZA-12
ZA-12/SiC/55_p
ZA-12/SiC/15_p

Between cast ZA-12 and inserts of:

Al-12wt%Si
Al-12wt%Si/SiC/55_p
ZA-12
ZA-12/SiC/55_p
ZA-12/SiC/15_p

Between cast Al-12wt%Si-1.5wt%Mg-1wt%Ni-1wt%Cu and inserts of:

Al-12wt%Si/SiC/55_p
ZA-12/SiC/55_p

Transient Liquid-Phase Bonding

Transient liquid-phase (TLP) bonding has proven fairly successful in joining Al-MMCs. The process employs a filler material or interlayer to produce a transient liquid layer at the interface to be joined. Solidification of the joint occurs isothermally, by diffusion. The advantage of TLP bonding of Al-MMCs is that it involves lower peak-processing temperatures than those used in other fusion welding processes. The lower processing temperatures result in less damage to the engineering properties of the composite in and around the joint. However, there is concern about damage to the properties of the composite base material during processing, because the thermal cycles for TLP bonding hold the entire assembly at temperatures near the solidus for a longer time than in the other fusion welding processes.

TLP bonding processes for Al-MMCs are controlled by several parameters, including surface finish, type and thickness of the interlayer or filler metal, processing time and temperature, and clamping pressure on the joint. The surface of Al_2O_3 must be removed to facilitate wetting of the substrate by the interlayer. The choice of the correct combination of processing parameters is critical in avoiding such problems as reinforcement-enriched or -depleted zones and Kirkendall porosity. In general, the thinnest interlayer possible should be used, to allow for the shortest processing times and a narrow liquid zone. Additionally, the lowest possible temperatures should be used, to limit damage to the material properties of the composite. Although the application of pressure can aid in improving interlayer-substrate contact, excess pressure can lead to expulsion of the liquid zone, which results in reinforcement-rich areas at the joint line.

SiC-Reinforced Al-MMC. TLP bonding with gold and Al-Si-Mg interlayers has been used to join two 6061 Al/SiC/25_p composite sheets (Ref 15). Based on thermomechanical simulation studies, a process window of time-temperature combinations was developed that would not produce damage to the properties of the base material. The window ranged from 30 min at 565 °C to 10 min at 580 °C. A 0.025 mm gold interlayer was shown to produce better results than the Al-Si-Mg material. Optimum conditions of 30 min at 567 to 580 °C were reported for the gold interlayer. Three distinct zones were observed in the vicinity of the bond interface: a particle-enriched zone, a diffuse flow zone that showed a pattern of material flow, and the undisturbed base material. No mention was made of the formation of Al_4C_3 for these conditions. Low tensile joint efficiencies (<30%) were attributed to voids at the joint interface.

Al_2O_3-Reinforced Al-MMC. The joining of 6061 Al/Al_2O_3/15_p using TLP bonding has also been reported (Ref 16). Three types of interlayers demonstrated an ability to join the material: gold, copper, and Al-12Si. The bonding tests were performed in vacuum (1.3×10^{-8}Pa) on mechanically cleaned parts using a clamping pressure of less than 70 kPa. Joining with a 25 μm thick interlayer of gold at 580 °C for 130 min produced the strongest bonds. The joints had a 323 MPa yield strength and a 341 MPa ultimate tensile strength with a joint efficiency of 95%. Joining with a 125 μm thick layer of Al-12Si at 585 °C for 20 min yielded bonds with nearly identical properties; however, the bond line contained residual filler material, as well as voids. A 25 μm thick layer of copper produced bonds with much lower strengths. An increase in volume fraction of the reinforcement was observed near the bond lines of joints made with gold and copper. Increased loading of reinforcement was not observed for bonds made with the Al-12Si.

Brazing

Brazing is an attractive alternative to other joining methods for attaching consolidated MMC preforms to the SPF/DB titanium structure. Brazing requires minimal bonding force and temperature exposure, eliminating or minimizing mechanical damage to fibers or fiber-matrix interactions.

Reference 17 reports the brazing of 6061 Al reinforced with 5, 10, and 15 vol% short Al_2O_3 fibers using Al-Si and Al-Mg interlayers. However, the reported bonding temperature range of 580 to 610 °C suggests that TLP bonding may have actually occurred. A 150 μm thick sheet of Al-10Si was used for joining, as well as a 140 μm thick sheet of Al-1.5Mn clad with 20 μm of Al-10Si on each side. The bonding time for each test was 10 min at a vacuum of 6.7 MPa. Bonding the composite to itself proved easier than joining the composite to monolithic 6061 Al. The tensile strengths for the similar material bonds of 5% Al_2O_3 fiber material exceeded 230 MPa.

Adhesive Bonding

Bergquist et al. (Ref 18) found that the application of conventional surface pretreatments to aluminum alloys reinforced with SiC_p produces surfaces that are quite different from those produced on bulk matrix metals. Grit blasting smears the surfaces with matrix metal, leaving them aluminum-rich. Standard etching treatments aggressively remove aluminum, produce surfaces enriched with loosely bound SiC_p, and change overall surface roughness. Examples are the FPL etch (Forest Products Lab, sodium dichromate/H_2SO_4) and the P2 etch (ferric sulfate plus H_2SO_4). Anodization limits SiC_p build-up and does not contribute to SiC enrichment. It was further found that the above changes depend on the time and type of the treatment. Three treated surfaces were selected for bonding at 121 °C with a commercial rubber-toughened epoxy film adhesive. Thick adherend lap shear bond strengths were found to be statistically independent of surface treatment and SiC_p reinforcement.

Very strong bonds, fully equivalent to if not better than those in bulk aluminum, can be formed in MMCs using standard bonding procedures and conventional film adhesives.

Introduction to CMCs

The use of ceramic monofilaments, whiskers, and/or particulates as reinforcement materials for ceramic matrices has led to the need to develop joining processes for the new CMC structural materials. The processes that have been investigated include hot press diffusion bonding, adhesive bonding, and brazing.

Hot Press Diffusion Bonding. The use of SiC monofilament to reinforce a Si_3N_4 matrix has led to refractory CMCs with outstanding high-temperature properties. These composites provide many of the properties required for future-generation aerospace applications and are viable candidates for structural applications where use temperatures exceed 1350 °C (Ref 19).

The fiber used as a reinforcement is a SiC monofilament produced by a high-temperature vapor deposition process using a carbon monofilament as a substrate. The vapor deposition process and the stoichiometric composition of the composites are responsible for the outstanding tensile properties (over 4 GPa at room temperature). Most studies were conducted with 140 μm fiber (SCS-6), and limited studies were conducted with 75 μm fiber (SCS-9). Both fibers, from Textron Specialty Materials, are products of vapor deposition of SiC on a carbon monofilament. The matrix is Si_3N_4 with 1.25% MgO and 5% Y_2O_3 as sintering aids.

Joining experiments with these materials have been conducted in anticipation of a need to fabricate complex shapes from an assembly of elementary geometric panels. For example, experiments evaluated a direct bonding approach wherein two densified composite panel specimens were hot pressed together using a monolithic monolayer as a high-temperature stable bonding agent. Double-notch shear tests showed that failure occurred outside the monolithic bond line region at levels of 140 MPa. No change in the interlaminar shear strength of the bonded specimens was observed after the specimens were exposed to thermal shock with a temperature gradient of up to 1200 °C.

Recent work has centered on hollow airfoil applications. To produce complex shapes such as hollow airfoils, it is necessary for the preform tape to conform to the curvature of the part and to achieve multidirection reinforcement. The tape contains a 75 μm diameter fiber (SCS-9), which results in easier wrapping of the tape around the leading edge of the airfoil. Other plies would be wrapped at opposite orientation angles and along the axis of the airfoil to give balanced and multidirectional reinforcement. Fabrication was performed by hot pressing. However, recent studies have centered on the use of HIP for consolidation of parts. Initial results indicate that HIP does not damage the fiber and produces a higher-strength composite.

A number of components have been made using the tape lay-up and hot press fabrication technique. They include vanes fabricated for an earth-to-orbit rocket engine turbine for General Electric under an ongoing NASA program.

Adhesive Bonding. Y-Si-Al-O-N is one of the most promising advanced engineering ceramics. These materials exhibit high strength and high toughness, and they show good wear and chemical resistance. At present, slip casting and injection molding are the main techniques for shaping, which limits the size and shape of the component that can be formed. The obvious alternative to these fabrication techniques is to use a joining method that can build up complex or large shapes from a series of smaller and more simply shaped components.

Techniques such as diffusion bonding, welding, soldering, or brazing cannot be directly applied to Si-Al-O-N ceramics because of their tendency to decompose rather than melt at high temperatures. However, Y-Si-Al-O-N ceramics contain a small percentage of intergranular glassy phase that melts at approximately 1350 °C (eutectic temperature in the Y-Si-Al-O-N system) (Ref 20). Hence, joining should be possible if this liquid-forming component of the ceramic is used.

The present joining adhesives selected differ from the pure-glass and composite adhesives in that they contain more solid phase than liquid phase at the joining temperature. The main constituent of the present adhesive is α-SiN (45 wt%), and the remainder is made up of a mixture of oxides (Y_2O_3, Al_2O_3, and SiO_2). The advantage of this type of adhesive is that it is closer in composition to the ceramic being joined, so the properties of the joint are not much different from those of the parent material. Also, at the joining temperature (1600 °C), the α-Si_3N_4 reacts to form β-SiAlON, which has an acicular nature that reinforces the joint. Material joined with pure-glass-forming adhesives is more likely to break at the joint, because the layer of glass between the adherends has low resistance to crack propagation. Figure 3-7 is a schematic representation of adhesive joining of Y-Si-Al-O-N ceramics.

Brazing. Understanding materials behavior during brazing of CMCs is an area of critical need. The broad class of materials serves a wide range of technologically important applications, including engine components, wear-resistant parts, metal extrusion dies, heat exchangers, machine tools, and medical products. Ceramic composites are often most easily introduced into an application by joining them to a metal support structure (Ref 21).

The active filler metal process has been developed to the point where reliable joints can be produced (Ref 22, 23). Silver-copper eutectics containing a small percentage of titanium and other minor alloying additions are among the most successful. The braze occurs by initial formation of a titanium-rich metal or metal oxide layer. Further transitions to the braze metal itself occur, as expected, via

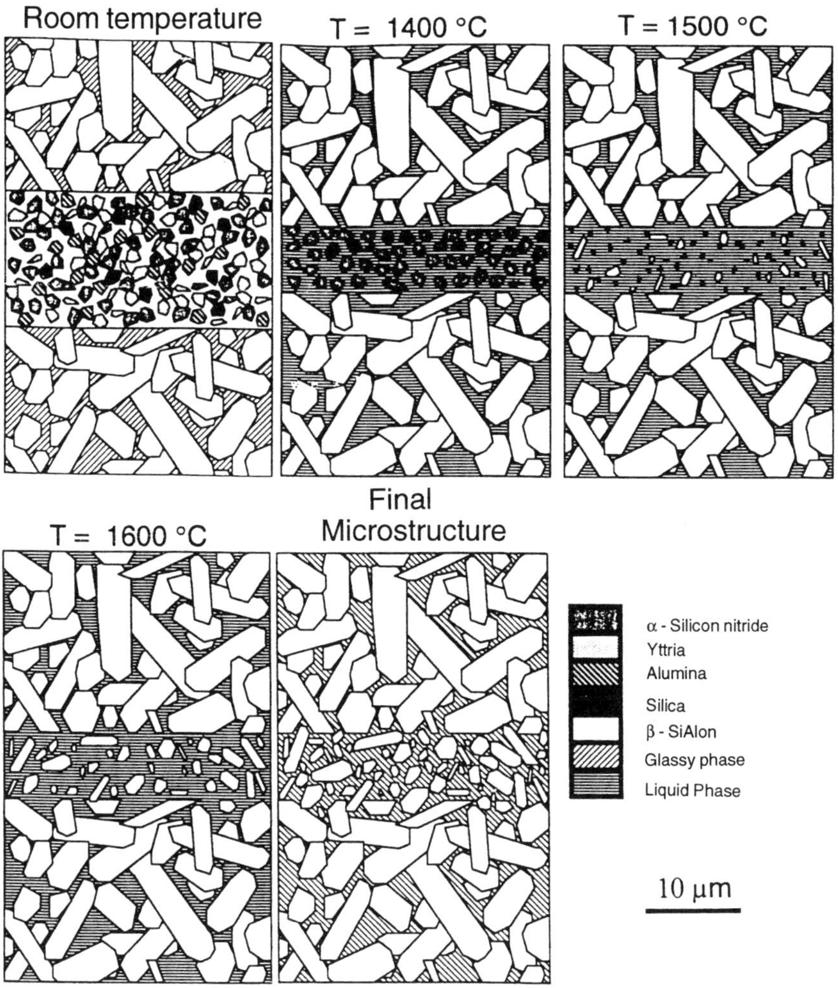

Fig. 3-7 Schematic representation of the adhesive joining process in Y-Si-Al-O-N ceramics. Source: Ref 20

gradients in both composition of the oxidation state. Other aspects of the joint involve the role of SiC whiskers at the interface.

A study on brazing of SiC_w/Al_2O_3 was performed by Lannutti et al. (Ref 21). They selected Incusil ABA as an active braze metal material, which consists of approximately 59.2 wt% Ag, 27.1 wt% Cu, 12.1 wt% In, and 1.2 wt% Ti. Transmission electron microscopy data suggested that the whiskers react extensively with the braze, that the silicon-base components then undergo transport away from their original site in the alumina matrix, and that the deposition of a continuous thin layer of these reaction products occurs away from the initial composite and braze reaction zone. Work continues toward:

- Identifying the mechanism responsible for the formation of observed discontinuities in the titanium-rich joint layer
- Investigating specific whisker-braze interactions, including their severity and net influence on joint strength
- Extending the investigation into other joining systems, such as nonoxide composites/oxide composites
- Correlating processing conditions, interlayer microstructure, and mechanical properties

Phosphate Bonding. Preparation of continuous fiber composites by hot pressing or HIP is expensive, and the high temperatures required may be detrimental to the fibers. The high temperatures may also create problems with the fiber-matrix interface and differential thermal expansion.

One way to prepare ceramic composites is to use phosphate bonding techniques. Phosphate binder systems have been used in common refractories (mostly alumina-based) for many years in the United States and abroad (Ref 24-29), but their use in high-performance ceramic composites, especially nonoxides, remains relatively unexplored. Aluminum phosphate has been studied as a binder for Si_3N_4 (Ref 30). Phosphate bonding techniques can provide a means for low-temperature processing of nonoxide ceramic composites, allowing exploitation of their unique properties (such as electrical) in applications such as electromagnetic windows where very high strength is not required.

Using phosphate bonding, CMCs can be prepared by well-developed techniques used in the PMC field. The general concept is the same: a continuous tow or cloth is combined with a liquid or paste matrix precursor, and with time and heat this matrix material solidifies. In the case of phosphate bonded ceramics, the matrix precursor is a suspension of ceramic particles in a liquid, instead of a monomer or oligomer. Nonoxide reinforcements (SiC_w and platelets and Si_3N_{4w}) introduced into aluminum phosphate bonded Si_3N_4 have been found to be chemically compatible with the binder.

The temperatures needed to cure phosphate binders are in the range of room temperature to 300 °C, not that different from the temperatures used in the polymer field. The lay-up and shape-forming methods used (vacuum bagging, pressing, autoclaving, etc.) are all applicable. In conventionally processed (sintered) composites, problems arise with densification because the reinforcement (fibers or whiskers) inhibits the shrinkage that must accompany sintering. Processing composites using phosphate bonding avoids these problems because linear changes are near zero, and practically any reinforcement loading can be used (Ref 31, 32).

For the continuous oxide fibers (silica and aluminosilicate) evaluated by Martin et al. (Ref 32), protective fiber coatings were necessary in order to prevent detrimental reactions with the matrix. Phenolic coatings were found to be effective only when composites were fired in nitrogen, effectively pyrolizing the phenolic. The resulting carbon coating on the fibers protected them from chemical interaction with the matrix. If this carbon coating had been detrimental to the dielectric properties, which are crucial for radome applications, it could have been removed by refiring the composites at 300 to 400 °C in an oxidizing atmosphere. Although strengths of continuous-fiber-reinforced composites in general were somewhat low, the highest strength observed (36 MPa for BPO_4/Si_3N_4 matrix, Astroquartz fused silica fabric) is adequate for radome applications, especially when accompanied by significant toughening of the material. The strength was very encouraging, considering that the material had 35% porosity.

The techniques for making continuous fiber composites are very similar to those used in the polymer composites field. Filament winding, infiltration of three-dimensional braids, tape wrap, and cloth lay-up techniques can all be used to produce phosphate bonded composites in net shapes. Another attractive feature of the phosphate bonding technique is negligible linear changes during firing, offering net-shape and net-size capabilities (Ref 31, 32).

In another study by Martin et al. (Ref 32), phosphate bonding techniques were used to prepare ceramics in the system Si_3N_4-Al_2O_3-B_2O_3-H_3PO_4 at low temperatures (700 to 900 °C). Mechanical properties and phase composition showed a strong dependence on alumina particle size and firing temperature. The use of 0.05 μm alumina resulted in stable phase composition and properties. The incorporation of boron into the binder system allows processing flexibility of materials before firing. The combination of strength, thermal expansion, and dielectric properties of phosphate bonded Si_3N_4 materials makes them promising candidates for applications such as advanced missile radomes and electromagnetic windows.

Carbon-Carbon and Graphite Joining

Various future aerospace systems, currently in the development stage, make use of structures and components that are fabricated from CCCs and monolithic graphite materials. These proposed systems use carbon-base materials for ground-

and space-based interceptor propulsion components, satellite structural components, space truss structures, and hypersonic vehicle propulsion and skin assemblies. In all cases, advanced joining methods need to be developed in order to fully exploit the capabilities of the CCCs and graphite materials and to obtain the performance characteristics desired. During the last half decade, technological advancements in CCCs and their product forms have made the issue of component assembly and joining methods ever more important.

Depending on the specific application, joints accomplished by brazing methods can have a number of advantages over mechanically fastened joints, including:

- The size and weight of brazed assemblies can be significantly less.
- Brazing permits the use of designs involving small, compact, multipart components.
- The use of high-temperature braze materials allows components with brazed joints to function at higher operating temperatures.

Components assembled by brazing methods have been designed primarily for various rocket propulsion applications, including hot gas valve component joints and injector/thrust-chamber joints. Joining schemes for assembly of CCC parts for the national aerospace plane have also been developed, as have joining schemes to allow use of CCC and graphite armor tiles in the large vacuum vessel/plasma chamber for the Doublet 3 fusion research device (Ref 33).

The various applications briefly discussed above represent a wide range of operating temperatures. Therefore, a number of different braze materials and techniques have been developed in order to achieve system design requirements and enhance the capabilities of the joined components. Some of the materials investigated for brazing CCC and graphite include silver-base filler metals, gold-base filler metals, zirconium metal, hafnium metal, and hafnium diboride/hafnium carbide powder mixtures. Other researchers have investigated brazing and diffusion bonding of CCC using titanium metal, silver-base filler metals, copper-base filler metals, various silicides ($MoSi_2$, BSi_2, and $TiSi_2$), and various proprietary materials (Ref 34-37).

For moderate-temperature applications, both silver- and gold-base braze filler metals have been used to join CCC panels to other CCC panels and to join graphite materials to refractory metals. Typical brazing temperatures have ranged from roughly 800 to 1250 °C for such applications. For high-temperature applications, zirconium braze materials have been used successfully with both CCC and graphite materials; typically, braze temperatures of approximately 1850 °C have been employed. Finally, for very-high-temperature applications, joints have been fabricated using both hafnium metal and HfB_2 powder braze materials to join both CCC and graphite materials; brazing temperatures of approximately 2200 to 2550 °C have been used. These braze material/operating temperature combinations are summarized in Table 3-2.

A study conducted by Dadras and Ngai (Ref 38) used brazing and solid-state diffusion bonding of three-dimensional CCC with Si, Si_3N_4, SiB_4, B, and $TiSi_2$ joining materials. The primary criteria for selecting these systems were:

- These materials and their carbides possess desirable high-temperature mechanical properties and thermal stability at temperature ranges greater than 1038 °C and less than 1928 °C.
- Elements with low atomic numbers are involved in these systems.
- Joining by solid-state diffusion bonding and carbide formation was expected to be possible.
- Joining temperatures below 2205 °C were expected.

The first three materials (Si, Si_3N_4, and SiB_4) produced no bonding or very weak bonds. Thus, it was determined that only the latter two systems (B and $TiSi_2$) were suitable for producing joints of sufficient shear strength ($\gamma > 3.4$ MPa). Due to the difficulty of melting the boron interlayers, only solid-state diffusion bonding with boron was studied.

The optimum conditions for joining were determined to be 1995 °C, 15 min, and 7.2 MPa. The shear strengths of the joints made under optimum conditions increased with the testing temperature from an average value of 5.8 MPa at room temperature to 18.1 MPa at 1660 °C. $TiSi_2$ exhibited excellent wettability and penetrability during the brazing operations. The optimum temperature for brazing was 1490 °C. The maximum shear strength of an optimum joint was 406 MPa at 1165 °C. The strength decreased at lower test temperatures, and at ambient temperature the average joint shear strength was 15.4 MPa (Ref 38).

Specialty Composite Joining

Functionally Gradient Materials (FGM). The application of FGM to the bonding of metals, ceramics, and composites is attracting attention from the viewpoint

Table 3-2 Carbon-carbon (C-C) and graphite joining overview

Primary substrates joined	Operating temperature	Braze material	Temperature for initiation of melting, °C
Graphite to refractory metal	Moderate	Silver, copper, titanium alloy	780
Graphite to refractory metal	Moderate	Gold, palladium alloy	1190
C-C to C-C, and graphite to graphite	High	Zirconium	1823
C-C to C-C, and graphite to graphite	Very high	Hafnium	2180
C-C to C-C, and graphite to graphite	Very high	HfB_2, with and without HfC	2260

Source: Ref 33

of reducing residual stress and eliminating intermediate materials and brazing. Development work has been limited to date. Nickel has been successfully joined to MgO with an FGM filler of Ni-NiO with good adhesion at the joining interfaces. The joined composite Ni/Ni-NiO(FGM)/MgO was formed by heating at 1300 °C at atmospheric pressure in air, and composition gradient was formed between the nickel and the solid solution.

An FGM filler of Ni-MgO was similarly used to join nickel to MgO, forming Ni/Ni-MgO(FGM)/MgO. This FGM was made by powder metallurgy. A compositionally gradient mixture of the metallic and ceramic powders was pressed (20 to 32 MPa) in a steel die followed by heating (627 to 1300 °C) in flowing nitrogen or air. The fracture strengths (three-point bending test) of the joints formed using the FGM Ni-NiO (72 to 128 MPa) and the FGM Ni-MgO (110 MPa) were higher than when nickel was directly joined to MgO (30 to 60 MPa).

Iron was also successfully joined to AlN via an FGM of Al-AlN by contacting the aluminum surface of the FGM with the iron and heating at 650 °C for 48 h in flowing argon. The microhardness of the aluminum-rich zone in the FGM was harder than that of pure aluminum. This solid-phase diffusion bonding technique provided a good Al-AlN FGM/pure iron junction. As SEM test results showed, the reactive layer was considered to be the $FeAl_3$ phase. Therefore, bonding with an FGM that has a metallic aluminum section made as thin as possible can add slight resistance against heat.

Aluminides. A new nonconventional friction stir welding process is a nonconsumable version of linear friction welding, wherein the rotating rod is consumed as it is moved across the workpiece. According to its developer, D. Nicholas of the Welding Institute, this new process has been used to weld titanium aluminides (Ref 39).

REFERENCES

1. M. Gittos and P. Threadgill, Joining Aluminum Alloy MMCs, Bulletin 5, *TWI J.*, Sept/Oct 1992, p 104-105
2. M.J. Cola, G. Martin, and C.E. Albright, "Inertia Friction Welding of Aluminum 6061/Al_2O_3 Particulates Reinforced Metal Matrix Composites," Edison Welding Institute Research Report Summary, 1989, p 7
3. B. Altshuller, W. Christy, and B. Wiskel, GMA Welding of Al-Al_2O_3 Metal Matrix Composite, *Weldability of Materials*, Proc. Materials Weldability Symp., ASM International, 1990, p 305-309
4. G.H. Reynolds, "Fusion Welding of Al/Al_2O_3 Composites—Phase I," MTL TR 90-38, DAAL04-89C0020, Final Rept., MSNW, Inc., Aug 1989-March 1990, July 1990
5. "The Weldability of Particle Reinforced Aluminum," Internal Report, Univ. of Wisconsin-Milwaukee, 1990

6. J.E. Gould, An Examination of Nugget Developments during Resistance Spot Welding Using Experimental and Analytical Techniques," *Weld. J.*, Vol 68 (No. 1), 1987, p 1s-10s
7. N.B. Dahotre, S. Gopinathan, M.H. McCay, et al., Laser Joining of Metal Matrix Composite, *Proc. Machining of Composite Materials Symp.*, ASM Materials Week (Chicago) 1-5 Nov 1992, p 167-173; TMS 1991, Ltwt. Alloys Aerospace Applica. II Mtg, p 313-325
8. S. Kawall and G.L. Viegelahn, Attempts at Laser Welding of an Alumina Reinforced 6061 Aluminum Alloy Composite, *Adv. Composite Materials: New Developments and Applications*, Conf. Proc. (Detroit), 30 Sept-3 Oct 1991, p 283-290
9. J.E. Hatch, *Aluminum-Properties and Physical Metallurgy*, American Society for Metals, 1984, p 359
10. T.J. Lienert, E.D. Brandon, and J.C. Lippold, Laser and Electron Beam Welding of SiC_p Reinforced Aluminum A356 Metal Matrix Composite, *Scr. Metall. Mater.*, Vol 28, June 1993, p 1341-1346
11. A. Cox, W.A. Baeslack, III, S. Zorko, et al., Capacitor-Discharge Resistance-Spot-Welding of SiC Fiber-Reinforced Ti-6Al-4V, *Weld. J.*, Vol 72 (No. 10), 1993, p 479s-491s
12. J.H. Devletian, SiC/Al Metal Matrix Composite Welding by a Capacitor Discharge Process, *Weld. J.*, Vol 66 (No. 1), p 33-39
13. R.H. Lorenz, T.E. Steelman, and W.D. Padian, "Selective Reinforcement of Low-Cost Titanium Components," NAAD, AFWAL-TR-80-4014, Final Report, Rockwell Int. Corp., Sept 1978-Oct 1979, Contr. F33615-78-C-5180, Air Force Materials Laboratory
14. T.E. Steelman et al., "Silicon Carbide/Titanium Material and Process Fundamentals," NAAD, IR-2420 (III), Interim Report Rockwell Int. Corp., 16 June 1980-15 Jan 1981, Contr. F33615-79-C-5012, Air Force Materials Laboratory
15. K. Sudhakar, "Joining of Aluminum Based Particulate-Reinforced Metal-Matrix Composites," Dissertation, Ohio State University, 1990
16. R. Klehn, "Joining of 6061 Aluminum Matrix Ceramic Particle Reinforced Composites," M.S. Thesis, Massachusetts Institute of Technology, Sept 1991
17. K. Suganuma, T. Okamoto, and N. Suzuki, Joining of Alumina Short-Fiber Reinforced AA6061 Alloy to AA6061 Alloy and to Itself, *J. Mater. Sci.*, 1987, p 1580-1584
18. P.R. Bergquist, S.P. Petrie, and S.E. Wentworth, An Assessment of the Adhesive Bondability of Silicon Carbide Reinforced Aluminum, *22nd Int. SAMPE Tech. Conf.*, L.D. Michelove, R.P. Caruso, and P. Adams, Ed., 6-8 Nov 1990, p 815-822
19. B. Thomson and J.-F. LeCostaouec, Recent Developments in SiC Monofilament Reinforced Si_3N_4 Composites, *SAMPE Q.*, Vol 22 (No. 3), April 1991, p 46-51
20. P.A. Walls and M. Ueki, Joining SiAlON Ceramics Using Composite β-SiAlON-Glass Adhesives, *J. Am. Ceram. Soc.*, Vol 75 (No. 9), 1992, p 2491-2497

21. J.J. Lannutti, E. Park, and J.D. Cawley, "Active Metal Brazing of Ceramic Matrix Composites," Edison Welding Institute, Res. Brief, B9201, Aug 1992
22. H. Mizuhara, E. Huebel, and T. Oyama, High-Reliability Joining of Ceramic to Metal, *J. Am Ceram. Soc.*, Vol 68 (No. 9), 1989, p 1591-1600
23. H. Mizuhara and K. Mally, Ceramic-to-Metal Joining Using an Active Filler Metal, *Weld. J.*, Vol 64 (No. 10), 1985, p 27-32
24. W.D. Kingery, Fundamental Study of Phosphate Bonding in Refractories: I, Literature Review, II, Cold Setting Properties, III, Phosphate Adsorption by Clay and Bond Migration, *J. Am. Ceram. Soc.*, Vol 33 (No. 8), 1950, p 239-250
25. F. Gonzalez and J. Halloran, Strength and Microstructure of Phosphate-Bonded Alumina Refractories, *Am. Ceram. Soc. Bull.*, Vol 60 (No. 7), 1981, p 700-702
26. M. Kuz'menkov, S. Plyshevskii, and I. Buraya, Interaction of an Aluminophosphate Binder with Some Fillers, *Obshch. Prikl. Khim.*, Vol 6, 1974, p 53-57
27. D. Karpinos et al., Physicochemical Processes Occurring in Nitride and Oxide-Nitride Compositions with Phosphate Binders during Heating, *Poroshk. Metall.*, Vol 5, 1982, p 50-54
28. F. Gonzalez and J. Halloran, Reaction of Orthophosphoric Acid with Several Forms of Aluminum Oxide, *Am. Ceram. Soc. Bull.*, Vol 59 (No. 7), 1980, p 727-738
29. F. Hummel, Properties of Some Substances Isostructural with Silica, *J. Am. Ceram. Soc.*, Vol 32 (No. 10), 1949, p 320-326
30. I.G. Talmy and C.A. Martin, Phosphate Bonded Non-Oxide Ceramics, *Proc. 14th Conference on Metal Matrix, Carbon and Ceramic Composites*, NASA CP-3097, Part 1, J.D. Buckley, Ed., 1990, p 309-316
31. C.A. Martin, I.G. Talmy, and A.H. Le, Phosphate Bonding in Processing of Whisker- and Fiber Reinforced Ceramic Composites, *Proc. 15th Conference on Metal Matrix, Carbon, and Ceramic Matrix Composites*, NASA CP-3133, Part I, J.D. Buckley, 1990, p 281-287
32. I.G. Talmy, C.A. Martin, A.H. Le, et al., Phosphate Bonding in Processing of Non-Oxide Ceramics, *Proc. 15th Conference on Metal Matrix, Carbon, and Ceramic Matrix Composites*, NASA CP-3133, Part I, J.D. Buckley, 1990, p 291-297
33. P.G. Valentine and P.W. Trester, Development of Brazed Joints in Carbon-Carbon, Graphite, and Refractory Metal Components for Rocket Propulsion and Spacecraft Applications, *Proc. 15th Conference on Metal Matrix, Carbon, and Ceramic Matrix Composites*, NASA CP-3133, Part I, J.D. Buckley, 1990, p 39-55
34. "Joining of Carbon-Carbon and Ceramic Matrix Composites," NSWC Presentation from the Interagency Planning Group Meeting, E. Becker, Ed.; IDA Memorandum Report M-312, T.F. Kearns, Ed., April 1987

35. "Joining of Carbon-Carbon and Ceramic Matrix Composites," Materials Innovation Labs Presentation from the Interagency Planning Group Meeting, S. Yalof, Ed.; IDA Memorandum Report M-312, T.F. Kearns, Ed.; April 1987
36. P. Dadras, Joining of Carbon-Carbon Composites by Using $MoSi_2$ and Titanium Interlayers, *Proc. 14th Conf. Metal Matrix, Carbon, and Ceramic Matrix Composites*, NASA CP 3097, Part 2; J.D. Buckley, Ed., 1990
37. P.G. Valentine and P.W. Trester, Reaction Sintering: A Method for Achieving Adherent High-Temperature Coatings on Carbon-Carbon Composites, *Proc. 15th Conf. on Metal Matrix, Carbon, and Ceramic Matrix Composites*, NASA CP-3133, Part II, J.D. Buckley, Ed., 1991, p 811-820
38. P. Dadras and T. Ngai, Joining of C-C Composites by Boron and Titanium Disilicide, *Proc. 15th Conf. Metal Matrix, Carbon, and Ceramic Matrix Composites*, NASA CP-3133, p 25-38, Part I, J.D. Buckley, Ed., 1991, p 25-38
39. *Weld. J.*, May 1993, p 18

SELECTED REFERENCES

- *Adhesive Bonding Handbook for Advanced Structural Materials*, ESA PSS 03-210, Issue 1, European Space Research and Technology Centre, Netherlands, Feb 1990
- N. Albee, Aspects of Mechanically Fastening Advanced Composites, *Adv. Compos.*, Nov-Dec 1988, p 54-60
- Aluminum Composites Improve Driveshafts, *Mater. Eng.*, April 1992, p 30
- "Assessment of Fatigue Damage in MMC and their Joints," Cooperative Research Project at Edison Welding Institute, Columbus, OH, Jan 1993, Vol 7 (No. 1), p 8
- Automation in Aerospace Fastening, *Aerosp. Eng.*, Aug 1992, p 13-18
- G. Bao, J.W. Hutchison, and R.M. McMeeking, "Particle Reinforcement of Ductile Matrices against Plastic Flow and Creep," Materials Dept., University of California, Santa Barbara, 1990
- M.A. Bayomi and M. Suery, "Structural and Mechanical Properties of SiC Particle Reinforced Al-Alloy Composites," Institut National Polytechnique de Grenoble, 1990
- D. Bhattacharyya, M.E. Bowie, and J.T. Gregory, The Influence of Alumina Microsphere Reinforcement on the Mechanical Behavior and Weldability of a 6061 Aluminum Metal Matrix Composite, *Proc. of the Mach. of Comp. Mater. Symp.*, ASM Materials Week (Chicago), 1-5 Nov 1992, p 49-56
- M.R. Bowditch and D.A. Moth, "Bonding Techniques with Particular Reinforced Metal Matrix Composites," British Defense Res. Agency, Sym. on Bonding of Advanced Composites, Adh. Grp. of British Institute of Materials, London, March 1993

- S. Brusethaug, "Mechanical Properties and Composition in A357 (Al17SiMg) PMMC," Hydro Aluminum N-Sunn-dalsora, 1990
- A.J. Caputo et al., Fiber-Reinforced SiC Composites with Improved Mechanical Properties, *Am. Ceram. Soc. Bull.*, Vol 66 (No. 2), 1987, p 368-372
- A.J. Caputo and W.J. Lackey, "Fabrication of Fiber-Reinforced Ceramic Composites by Chemical Vapor Infiltration," Tech. Memo 9235, Oak Ridge National Laboratory, 1984
- "Carbon-Fiber-Reinforced Solder Preform," Patent No. 5,089,356
- J. Carroll and S. Moran, "A Review of Lockheed-Georgia's R&D Programs in Aluminum Matrix Composites," 17th Nat. SAMPE Tech. Conf. (Kiamesha Lake, NY), 22-24 Oct 1985
- A.B. Carter, III, Preserving the Structural Integrity of Advanced Composite Materials through the Use of Surface Mounted Fasteners, *SAMPE J.*, Vol 25 (No. 4), July-Aug 1989, p 21-25
- V.M. Cassidy, Composite Metal Auto Parts Only Five Years Away, *Mod. Met.*, Dec 1989, p 62-66
- J. Cao and D.D.L. Chung, Carbon Fiber Silver-Copper Brazing Filler Composites for Brazing Ceramics, *Weld. J.*, Vol 71 (No. 1), 1992, p 21s-24s
- M.J. Cola, "Inertia-Friction Welding of Particulate-Reinforced Aluminum Matrix Composites," Thesis, Ohio State University, 1992
- M. Cola, W.A. Baeslack, III, and B. Altshuller, Paper E, Session 3, 12 April 1994, 75th Ann. Inter. Conv. (Philadelphia), American Welding Society
- M.J. Cola, T.J. Lienert, J.E. Gould, et al., Laser Welding of a SiC/Al-Alloy Metal Matrix Composite, *J. Appl. Phys.*, Vol 65 (No. 12), 1989, p 5072-5077
- M.J. Cola, G.S. Martin, and C.E. Albright, "Inertia Friction Welding of a 6061-T6/Al_2O_3/15_p Metal-Matrix Composite," Research Rept. MR9108, Edison Welding Institute, June 1991
- P. Dadras and G. Mehrotra, "Solid-State Diffusion Bonding of Carbon-Carbon Composites with Borides and Carbides," Rept. MR 9206, EWI, June 1992
- N.B. Dahotre, M.H. McCay, T.D. McCay, et al., Laser Joining of Metal Matrix Composite, *Proc. Machining of Composite Materials Symp.*, ASM Materials Week (Chicago), 1-5 Nov 1992
- N.B. Dahotre, M.H. McCay, T.D. McCay, et al., Laser Processing of a SiC/Al-Alloy Metal Matrix Composite, *J. Mater. Sci.*, Vol 6 (No. 3), 1991, p 514-529
- N.B. Dahotre, T.D. McCay, and M.H. McCay, Laser Processing of a SiC/Al-Alloy Metal Matrix Composite, *J. Appl. Phys.*, Vol 65 (No. 12), 1989, p 5072-5077
- S.A. David, D.N. Braski, and C.T. Liu, Structure and Properties of Welded Long-Range-Ordered Alloys, *Weld. J.*, Vol 65 (No. 4), 1986, p 93-s
- S.A. David, J.A. Horton, C.G. McKamey, et al., Welding of Iron Aluminides, *Weld. J.*, Vol 68 (No. 9), 1989, p 172-s

- S.A. David, W.A. Jemian, C. Liu, et al., Welding and Weldability of Nickel-Iron Aluminides, *Weld. J.*, Vol 64 (No. 1), 1985, p 22-s
- S.A. David and M.L. Santella, Welding Behavior and Welding Metallurgy of Ductile Intermetallic Alloys, *Advances in Welding Metallurgy*, Proc. of the 1st U.S.-Japan Symp. (Miami), American Welding Society, 1990, p 65
- S.A. David and T. Zacharia, Weldability of Fe_3Al-Type Aluminide, *Weld. J.*, Vol 72 (No. 5), May 1993, p 201-s to 207-s
- R.J. Davies and M.D. Ritchie, Future Design Concepts for the Development of New Pretreatments of Aluminum Alloy/Metal Matrix Composite for Adhesive Bonding, *J. Adhes.*, Vol 38, 1992, p 243-254
- D.V. Dunford, S.M. Flitcroft, D.S. McDarmaid, et al., Forming and Diffusion Bonding of Al-Li Alloy Containing 20 wt% Particulate SiC, *Aluminum-Lithium*, M. Peters and P.-J. Winkler, Ed., Vol 2, DGM, Informationsgesellschaft-Verlag, 1992, p 1087-1092
- Fasteners for Composites, *Aircr. Eng.*, July 1988, p 2, 3
- S.A. Gedeon, R. Guerriero, and I. Tangerini, "Process for Obtaining a Metallurgical Bond between a Metal Material, or a Composite Material Having a Metal Matrix and a Metal Casting or a Metal-Alloy Casting," European Patent 89202324.3-, Oct 1989
- S.A. Gedeon, C. Lane, and B. Altshuller, Welding Institute of Canada, Paper C, Session 16, 75th Ann. Inter. Conv. (Philadelphia), 14 April 1994, American Welding Society
- S.A. Gedeon and I. Tangerini, A New Method for Bonding Metal Matrix Composite Inserts during Casting, *Mater. Sci. Eng. A*, Vol 144, 1991, p 237-241
- D.M. Ginburg, "Metal Matrix Composite Vertical Tail Fabrication," EM90-438, Metal Matrix Clinic (Anaheim, CA), 13-14 Nov 1990, Society of Manufacturing Engineers
- M. Gittos and P. Threadgill, Joining Aluminum Alloy MMCs, Bulletin 5, *TWI J.*, Sept-Oct 1992, p 9-10
- A. Hirose, M. Kotoh, S. Fukumoto, et al., Diffusion Bonding of SiC Fibre Reinforced Ti-6Al-4V, *Mater. Sci. Technol.*, Vol 8 (No. 9), Sept 1992, p 811-815
- A. Hirose, Y. Matsuhiro, M. Kotoh, et al., Laser-Beam Welding of SiC Fibre-Reinforced Ti-6Al-4V Composite, *J. Mater. Sci.*, Vol 28, 1993, p 349-355
- E.K. Hoffman, R.K. Bird, and D.L. Dicus, Session B3, Paper B3A, Brazed Joint Properties and Microstructures of β21S Titanium Matrix Composites, *Composite Materials Technology*, Hanser Publishers, 1992
- D. Hughes, Textron Applies Automation to Titanium Matrix Process, *Aviat. Week Space Technol.*, 15 June 1992, p 87-88
- B. Irving, What's Being Done to Weld Metal-Matrix Composites? *Weld. J.*, Vol 70 (No. 6), June 1991, p 65-67

- T. Iseki, T. Kameda, and T. Maruyama, Interfacial Reactions between SiC and Aluminum during Joining, *J. Mater. Sci.*, Vol 19, 1984, p 1692-1698
- "Joining Metals to Ceramics Using FGMs, Materials and Processing Report," Elsevier Science, March/April 1992, p 5, 6
- S.M. Kawali, G.L. Viegelahn, and R. Scheuerman, Laser Welding of Alumina Reinforced 6061 Aluminum Alloy Composite, *Proc. Laser Materials Processing Symp.*, ICALEO '91, Laser Institute of America, 1991, p 156-167
- A. Lacombe and C. Bonnet, "Ceramic Matrix Composites, Key Materials for Future Space Plane Technologies," AIAA-90-5208, American Institute of Aeronautics and Astronautics, Oct 1990
- R. Larker, B. Loberg, and T. Johansson, *Proc. 3rd Int. Symp. on Ceramic Materials and Components for Engines* (Las Vegas, NV), V.J. Tennery, Ed., American Ceramic Society, 1989, p 503-512
- R. Larker, A. Nissen, L. Pejryd, et al., Diffusion Bonding Reactions between a SiC/SiC Composite and Two Superalloys during Joining by Hot Isostatic Pressing, *Acta Metall. Mater.*, Vol 40 (No. 11), 1992, p 3129-3139
- R. Larker, L.-Y. Wei, M. Olsson, et al., Diffusion Bonding of Si_3N_4/TiN and Si_3N_4/TiB_2 Composites to Incoloy 909, *Proc. of 4th Int. Symp. on Ceramic Materials and Components for Engines*, R. Carlsson, T. Johansson, and L. Kahlman, Ed., American Ceramic Society, Elsevier Applied Science, 1992, p 340-347
- S.C. Levin and V.P. McConnell, Two-Wheeler Evolution, *Adv. Compos.*, Jan/Feb 1993, p 39-45
- T.J. Lienert, W.A. Baeslack, III, M. Kuo, et al., "Inertia Friction Welding of SiC-Reinforced 8009 Aluminum," Paper A, Session 16, 75th Ann. Inter. Conv. (Philadelphia), 14 April 1994, American Welding Society
- C. Liu, S. Pape, and J.J. Lewandowski, *Interfaces in Polymer, Ceramic, and Metal Matrix Composites*, H. Ishida, Ed., Elsevier, 1988
- D.J. Lloyd, *Compos. Sci. Technol.*, Vol 35, 1989, p 159
- D.J. Lloyd, Metal Matrix Composite—An Overview, *Advanced Structural Materials* (Montreal), 28-31 Aug 1988, D.S. Wilkinson, Ed., Vol 9, Pergamon Press, 1989, p 1-20
- S.H. Lo, S. Dionne, M. Popescu, et al., Effects of Prior- and Post-Weld Heat Treatments on the Properties of SiC/AlSi Composite Gas Metal Arc Welded Joints, *Proc. Advances in Production and Fabrication of Light Metals and MMCs*, M. Avedisian, L. Larouche, and J. Masounave, Ed., Canadian Institute of Mining, Metallurgy and Petroleum, 1992, p 575-588
- Lockheed-Georgia Begins Design of Metal Matrix Composite Tail Fins, *Aviat. Week Space Technol.*, 18 Nov 1985, p 127
- M. Mabuchi, T. Imai, K. Kubo, et al., New Fabrication Procedure for Superplastic Aluminum Composites Reinforced with Si_3N_4 Whiskers or Particulates, *7th Ann. ASM/ESD Advanced Composites Conf. and Expo.* (Detroit), 30 Sept-3 Oct 1991, p 275-282

- P.K. Mallick and S. Newman, Ed., Section III, Chapter 6A, *Composite Materials Technology*, Hanser Publishers, 1992, p 179-210
- Manufacturers are Developing Improved Structural Systems, *Aviat. Week Space Technol.*, 12 Dec 1988, p 67-72
- R.P. Martukanitz and R.B. Bhagat, Laser Processing of Discontinuously Reinforced Aluminum Matrix Composites, *The Metal Science of Joining*, ASM International, 1992, p 241-247
- L.H. McCarty, Lightweight Fasteners Reduce Unit Stresses on Composites, *Des. News*, 4 Sept 1989, p 186, 187
- G.E. Metzger, "Joining of Metal-Matrix Fiber-Reinforced Composite Materials," Bull. 207, Welding Research Council, 1975
- O.T. Midling, "Friction Welding of SiC Particle Reinforced Aluminum Alloy," STF-34-A-91030, Selskapet for Industriell Og Teknisk Forskning, Norway, April 1991
- M.R. Morel, D.A. Saravanos, and C.C. Chamis, "Tailored Metal Matrix Composites for High-Temperature Performance," TM-105816, E-7247, National Aeronautics and Space Administration, March 1992
- T. Nagano and H. Kato, Diffusion Bonding of Zirconia/Alumina Composites, *J. Am. Ceram. Soc.*, Vol 73 (No. 1), 1990, p 3476-3480
- P.G. Partridge and D.V. Dunford, Interlayers and Interfaces in Diffusion Bonded Joints in Metal Matrix Composites, *Aluminum-Lithium*, M. Peters and P.-J. Winkler, Ed., Vol 2, DGM, Informationsgesellschaft-Verlag, 1992, p 1145-1150
- P.G. Partridge and D.V. Dunford, The Role of Interlayers in Diffusion Bonded Joints in Metal-Matrix Composites, *J. Mater. Sci.*, Vol 26, 1991, p 2255-2258
- P.G. Partridge, M. Shepherd, and D.V. Dunford, Statistical Analysis of Particulate Interface Lengths in Diffusion Bonded Joints in a Metal-Matrix Composite, *J. Mater. Sci.*, Vol 26, 1991, p 4953-4960
- A.V. Pocius and R.P. Wenz, *SAMPE J.*, Vol 21 (No. 5), 1985, p 50
- B.H. Rabin, Joining of Fiber-Reinforced SiC Composites by in Situ Reaction Methods, *Mater. Sci. Eng., A*, Vol 130, 1990, p L1-L5
- B.H. Rabin, Joining of Silicon Carbide/Silicon Carbide Composites and Dense Silicon Carbide Using Combustion Reactions in the Titanium-Carbon-Nickel System, *J. Am. Ceram. Soc.*, Vol 75 (No. 1), 1992, p 131-135
- J. Rausch, Fastening Fundamentals for Fabricated Composites, *Assembly Eng.*, Oct 1988, p 34-36
- D.H. Rehbein, I. Decker, and H. Wohlfahrt, "Laser Beam Welding of Aluminum Die Castings with Reduced Pore Formation," Paper E, 75th Ann. Inter. Conv. (Philadelphia), 14 April 1994, American Welding Society
- P.K. Rohatgi, R. Asthana, and S. Das, Solidification, Structures, and Properties of Cast Metal-Ceramic Particle Composites, *Int. Met. Rev.*, Vol 31 (No. 3), 1986, p 115-139

- S.N. Rosenwasser, A.J. Auvil, and R.D. Stevenson, "Development of Low-Temperature, Solid-State Bonding Approach for Metal Matrix Composite Joints," TR 89-302, LJ 89-040-TR, N60921-86-C-0279, NSWC, Oct 1989
- M.L. Santella and S.A. David, Weldability of Ni_3Al type Aluminide Alloys, *Weld. J.*, Vol 65 (No. 5), 1986, p 129-s
- R. Schonholz, G. Kleer, and W. Doll, *Reliability of Interfaces in Newly Designed Ceramic-Ceramic and Metal-Ceramic Systems*, Fraunhofer-Institut fur Werkstoffmechanik, Freiburg, Germany, July 1990
- D.S. Schwartz, J.H. Devletian, S.J. Chen, et al., Capacitor Discharge and Flash Welding of SiC-Reinforced High Temperature Al Alloys, *Joining and Adhesion of Inorganic Materials*, Vol 314, Proc. Spring 1993 Mtg. Materials Research Society
- M.M. Schwartz, *Composite Materials Handbook*, 2nd ed., McGraw-Hill, 1992
- J.P. Sorensen, "Titanium Matrix Composites-NASP Materials and Structures Augmentation Program," AIAA 90-5207, American Institute of Aeronautics and Astronautics, Oct 1990
- M.J. Strum and G.A. Henshall, "Diffusion Bonding of NiAl with Ni and Al Interlayers," Paper C, Session 3, 12 April 1994, 75th Ann. Inter. Conv. (Philadelphia), American Welding Society
- P.E. Teague, Fasteners that Mate with New Materials, *Des. News*, 21 Sept 1992, p 98-102
- Textron Plant Makes Titanium MMCs, *Adv. Mater. Proc.*, Vol 141 (No. 4), April 1992, p 9
- Thermoplastics Inch Ahead, *Plast. Technol.*, March 1992, p 44-45
- "Titanium Matrix Composites," NASP Materials and Structures Augmentation Program, McDonnell Douglas Corp., Contract No., F33657-86-C-2126
- L. Trego, Fasteners for Aerospace Structures, *Aerosp. Eng.*, Nov 1991, p 19-23
- J.A. Vaccari, Processing Metal-Matrix Composites, *Am. Mach.*, March 1985, p 95-97
- F. Wakai and H. Kato, Superplasticity of TZP/Al_2O_3 Composite, *Adv. Ceram. Mater.*, Vol 3 (No. 1), 1988, p 71-76
- F. Wakai, Y. Kodama, S. Sakaguchi, et al., Superplastic Deformation of ZrO_2/Al_2O_3 Duplex Composite, *Proc. of the MRS Int. Mtg. on Advanced Materials*, Vol 7, *Superplasticity*, M. Kobayashi and F. Wakai, Ed., Materials Research Society, 1989, p 259-266
- F. Wakai, S. Sakaguchi, and Y. Masuno, Superplasticity of Yttria-Stabilized Tetragonal ZrO_2 Polycrystals, *Adv. Ceram. Mater.*, Vol 1 (No. 3), 1986, p 259-263
- M. Zhu and D.D.L. Chung, Active Brazing Alloy Containing Carbon Fibers for Metal-Ceramic Joining, Session B3, Paper B3B, *Composite Materials Technology*, Hanser Publishers, 1992

Chapter 4

The Future for Composites

Demand for higher-quality products, as reflected by stricter specifications for processing materials, is the principal factor driving the continuing improvements being made in composites and their hybrid variations. Energy concerns, environmental impact, new materials use, and advancement of new, vigorous manufacturing processes are other influential factors, all interrelated, that are governing the direction of industrial activities today and will continue to do so in the foreseeable future.

Resin Matrix Composites and Polymer Matrix Composites

The trend in PMC technology is toward materials with improved performance, such as increased strength, increased stiffness, higher-temperature capability, and improved environmental resistance. To meet the performance improvements goals, new fiber sizes, shapes, and chemical compositions are the trend in reinforcements. For improved performance in resins, new chemical formulations are the trend. To improve the affordability and producibility of PMC hardware, the trend is toward new material forms, new processing methods, new manufacturing techniques, and inspection methods that employ automation.

Currently, the most significant trend is aimed at improved performance, affordability, and producibility of thermoplastic resin systems. Thermoplastic resin systems offer improved durability/toughness, reduced processing time, and unique product forms. Figures 4-1 and 4-2, which were published in 1989, predict 15-year trends.

Joining Technology

As another way to improve the performance of the PMC structure, there is a trend toward improved joining technology, including both improved fasteners for mechanical joints and new adhesives for bonded joints.

Parameter	Trend
Fibers	Increased strain to failure, PBT fibers, metal-coatcd fibers, noncircular libers
Resins	Thermoplastics, IPNs, copolymers, LCPs, conductive polymers, fiber sizings, HT resins, HT adhesives, tough resins, ordered polymers, moisture-resistant resins
Metal Forms	Molding compounds, woven preforms, thermoplastic sheets, comingled forms, unidirectional tape/fabric
Processing/ Manufacturing	CAM/CIM, low-cost methods, thermo-forming; fabrication in space; CAP, scale up to large aircraft
Quality Assurance	CAI/CII, integral surface damage indicators, real-time imaging; internal sensors, FAI methods

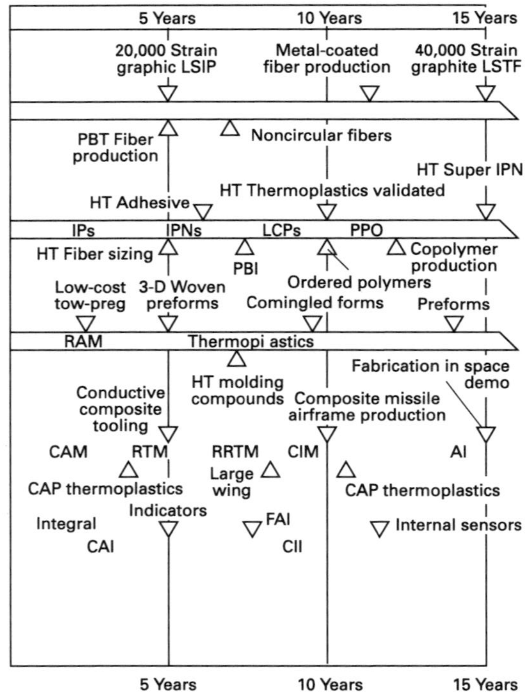

Fig. 4-1 Forecasts for polymer matrix composites. AI, artificial intelligence. CAI, computer-assisted inspection. CAM, computer-integrated manufacturing. CAP, computer-assisted inspection. FAI, frequency amplitude inflection inspection. HT, high-temperature. IPN, interpenetrating polymer network. LCPs, liquid crystal polymers. LSTF, laminate strain to failure. PBI, polybenzimidazole. PBT, polybutyleneterphthalate. PPQ, polyphenylquinoxaline, RAM, radar-absorbing material. RRIM, reinforced reaction injection molding. RTM, resin transfer molding. Source: Ref 1

Parameter	Trend
Joints	Improved mechanical fasteners, HT adhesives, TP joining methods
Repair	Post-buckled structure, TP repair, RAM/RAS repair, FW structure repair, repair in space, BDR, joints, rapid-cure adhesives
Design/Analysis	Computer integrated analysis, design optimization routines
Performance Demonstrations	Fatigue, thermal cycling, high-energy laser, low-velocity impact, rain/dust erosion, microparticle bombardment, nuclear/biological/chemical, radar cross section, thermal pulse, space debris bombardment
Supportability/ Maintainability	Noninspectable structures, resistance to in-service damage caused by maintenance activity

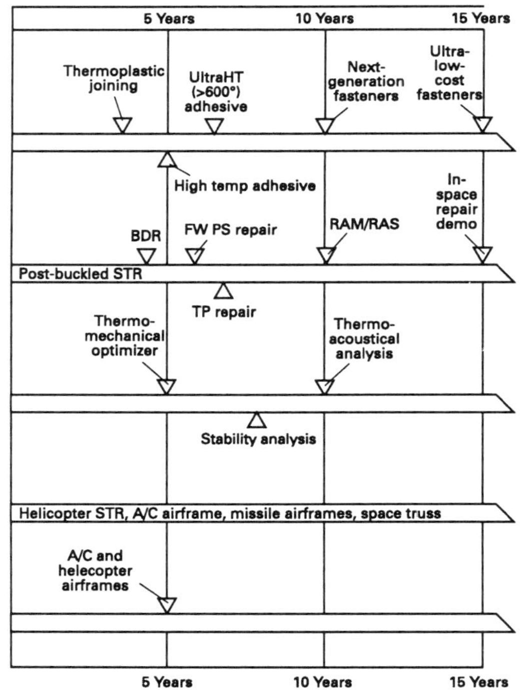

Fig. 4-2 Forecasts for polymer matrix composites. A/C, aircraft. BDR, battle damage repair. FWPS, filament would primary structure. HT, high-temperature. RAM, radar-absorbing material. RAS, radar-absorbing structure. STR, structure. TP, thermoplastic. Source: Ref 1

Bolts and Other Mechanical Fasteners. Some of the considerations in mechanical fastening are material selection, the brittleness of composites, and ensuring the strength of the laminate against shear and bending stresses in the fasteners.

Materials Selection. Metallic fasteners have succeeded very well with fiberglass, boron, and Devlar fiber composites. However, graphite reinforcements present problems. Some metallic fasteners in contact with graphite composites are prone to corrosion. Aluminum is the worst in this respect; stainless steel is somewhat better. Nickel and titanium alloys show excellent compatibility with graphite but are expensive. This kind of corrosion especially concerns aircraft designers because of their growing use of graphite. Other than relying on the more expensive metal fasteners, two main solutions are possible:

- Insulate the graphite from the metal by surrounding an aluminum or steel fastener with fiberglass or adhesive with scrim. (Scrim is a low-cost reinforcing fabric made from continuous filament yarn in an open-mesh construction.) Although this approach prevents corrosion, it complicates fastener installation.
- Use composite fasteners rather than metallic ones. Glass-epoxy and graphite-epoxy do not corrode galvanically, and they save weight. A graphite-epoxy fastener is five times lighter than a stainless steel fastener of similar size, and glass-epoxy is four times lighter. Tests by one helicopter manufacturer indicate that fiberglass-composite fasteners are sound for light to medium loads, although they are somewhat degraded by humidity.

In the end, however, design loads and stresses dictate the choice of fasteners. A high-performance aircraft structure may require titanium fasteners despite the cost.

Brittleness. Most composites are quite brittle, which means that their strains to failure are low. Thus, stress concentrations (and strains) caused by bolts themselves can lead relatively easily to failure of the surrounding composite.

One way to understand this is to compare composites with metals. Suppose that a bolt hole in aluminum or steel is slightly too small. When the bolt is inserted, stresses arise in the surrounding metal. If these stresses are great enough, the steel or aluminum yields, shifting some of the load to adjacent bolts. At the same time, stress is reduced around the original hole. Brittle composites, on the other hand, do not have yield points. If one bolt is tighter in its hole than others, the bearing stress at that hole simply remains higher. If loading increases this stress beyond the ultimate stress, the composite does not yield, but rather fails. All of the load will then suddenly fall on adjacent bolts, perhaps causing failure there as well. Figure 4-3 illustrates these differences between metal and composite bolt-stress behavior.

One way to alleviate this problem is to "soften" the composite in the bolt area by using staggered fabric plies to reduce brittleness. Another approach is to coat the bolt with uncured resin before inserting it. The resin flows into any open areas between the bolt and the composite. When cured, it helps to distribute loads more

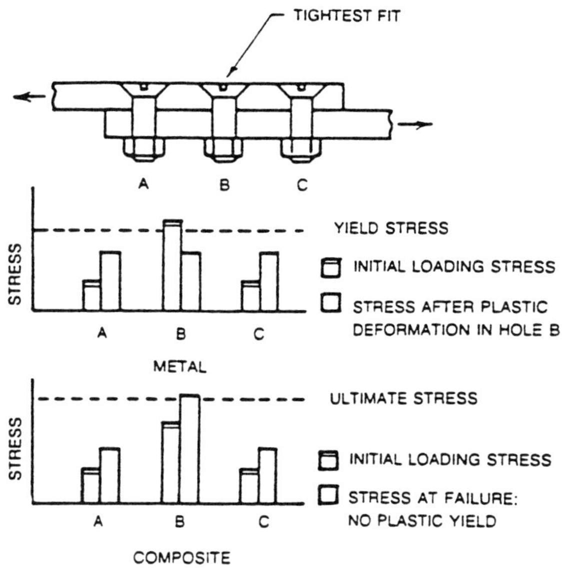

Fig. 4-3 Effect of bolt tightness stresses in metals and composites

evenly. The cured resin would also be less brittle than surrounding fibers, such as graphite. Finally, bonded metal inserts can be used, so that the bolt is actually in the metal, not the composite. However, this approach is costly and adds weight.

One way *not* to attack the problem is to reinforce the hole with more high-modulus fibers, such as graphite. This only increases brittleness around the hole, making the joint more likely to fail under bolt-induced stresses.

Laminate Strength versus Stresses in the Fasteners. As in metallic joints, prime considerations in composite joint design are bearing, tensile, and compressive strengths in the laminate, and shear and bending stresses in the fasteners.

Advanced composites can be delicate materials to fasten. Ordinary fasteners are not designed to protect the fibrous panels and work within the strength limitations. These materials are sensitive to radial expansion of the holes, which creates undue stresses, and to pressure on the back or blind side, which can result in fastener pull-through. The pressure required during installation can cause buckling or delamination, and the problems that can occur during use include delamination, crushing, and pull-through (Ref 2-8).

Composites require special treatment for several reasons. Because they have lower thickness compression strength than metals, the blind fastener head should expand more for proper fastening. Too small a clamp area can cause excessive load on the bearing area, resulting in crushing or delamination around the hole. The pull-up length of the fastener should precisely match the thickness of the panels being joined, to avoid local crushing. The fastener should not expand within the hole, because that also causes delamination. A larger clamp area may prevent pull-through. In addition, holes must be drilled cleanly, minimizing splintering and fraying and consequent fiber separation (Ref 8).

To control compressive stresses, engineers have designed various fasteners with large bearing areas to distribute the load over a large area. Installation problems have also been reduced or eliminated through customized fastener designs and the use of special installation tools.

Adhesive Joining. The examples below illustrate the new designs that are possible with improved adhesives.

Sailboat Construction. With the development of advanced composite process (ACP) technology, Advance USA of East Haddam, CT has created a hybrid composite for use in the construction of 4 m sailboats in lieu of fiberglass. ACP consists of a coextruded thermoplastic outer skin, reinforced with a urethane foam and glass fiber superstructure to form a high-strength "sandwich." This produces a boat that weighs less and is more rigid than a fiberglass design.

In the development of ACP technology, concurrent adhesives were required to securely bond the thermoplastic decks and hulls. A Plexus adhesive (nonstyrenated methacrylate) was found that had the same flexibility as the thermoplastic material itself, providing high-elongation, flexible joints as opposed to the brittle joints that conventional systems provide. No surface preparation was required, and because the adhesive components were combined in 1:1 ratios, they were much easier to mix and disperse.

The main reason that Advance USA uses a thermoplastic composite is its relative economy compared to fiberglass. The fiberglass design also required sandwich construction, but in fiberglass this type of construction is prohibitively expensive. ACP construction also reduces labor costs compared to fiberglass production. The adhesive allows engineers to design without the need for conventional fasteners, cures at room temperature without the need for heating or UV radiation, requires no primer or surface preparation, dispenses easily, and possesses superior gap-filling characteristics.

Armored Vehicles. Thick laminated composite hulls for armored vehicles are being developed to provide lighter, quieter vehicles that will meet all service requirements. Tests are conducted to determine data for use in finite element models for both adhesives and composites (Ref 9). A high-performance adhesive was selected to withstand a load that, in nearly all tests, caused failure in the composites before the adhesives failed. The joint design was expected to meet all requirements.

Testing showed that two joint designs could meet the goals for thick composite laminate vehicle structures: a scarf joint and a butt joint (Fig. 4-4a, b). The scarf joint was the better of the two designs, based on the analytical results using the proposed criterion. Ultimate failure occurred at a load about 1.5% higher than that at the butt

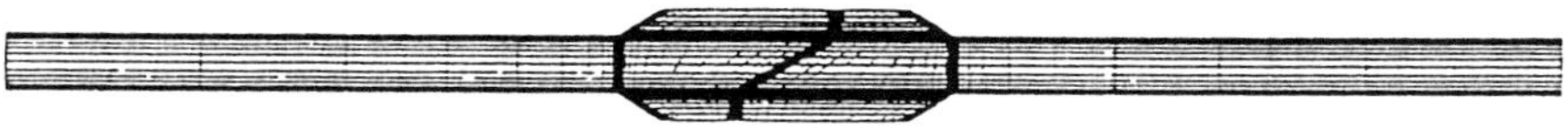

(a)

(b)

Fig. 4-4 Finite element model of joints. (a) Scarf joint. (b) Butt joint. Source: Ref 9

joint. The analysis method predicted failure initiation at 1.45% the butt joint load capability. As predicted, laminate failure was the actual failure mode in these joints.

Welding. The developments in thermoplastics have led to the need to assess microstructural changes in welds, examine alternate welding processes, and develop new processes tailored to these materials. Additionally, rapid thermal cycle processes are being investigated to minimize particle-matrix interactions with regard to particulate-reinforced aluminum composites. The concern has been that during fusion welding (particularly with Al-SiC composites), the SiC particles break down in the superheated molten aluminum, allowing the subsequent precipitation of a range of detrimental phases on cooling. Large aluminum carbide plates dominate the structure. Scientists and welding engineers have postulated which mechanisms create these changes and have attempted to predict which welding processes would be most applicable to these types of materials.

Solid-state processes have been considered advantageous because there is no melting to potentially allow dissolution of the reinforcing particles. The resulting structure shows essentially a base metal distribution of MMC particles all the way to the bond line. As for continuously reinforced titanium materials, considerable work has been done to develop solid-state variations of existing rapid thermal cycle processes, including resistance spot diffusion bonding and resistance brazing. These approaches have the advantage of achieving a bond at the contacting surface and yet avoiding the high heat that would degrade the reinforcing fibers. Application of other processes, including ultrasonic welding and resistance implant welding, are also under consideration by several firms.

Repair

As PMCs are introduced into service in aerospace systems, an emphasis on the operational supportability of the composite hardware is developing. Repair technology for all types of composite structures is a primary focus of technology development.

Some damage to composite structures, particularly in aircraft, is not visible but is of concern. Because composites lack strain capability, the damage to a composite structure may be more severe than that incurred by a comparable metal structure under an identical impact. Whereas the metal may yield to show the impact site, the composite may delaminate within and not reveal the damaged area. In-plane tensile strength may not always be compromised by impact damage to the matrix, but a damaged matrix cannot stabilize fibers under compressive bending or shear stresses, and compression is usually the critical loading mode in aircraft structures.

The steps in most repair maintenance programs are to:

1. Investigate and quantify the extent of damage.
2. Determine how comprehensive a repair is needed and where the repairs will be performed. The options are: (a) Repair to full properties (with all flaws detected and corrected). (b) Repair to acceptable properties, which will allow the structure to be usable to some high, predetermined percentage of full-scale operation. (c) Repair to some emergency level that will allow use at a low percentage of operational effectiveness (e.g., fly damaged aircraft to depot maintenance base).
3. Select the repair configuration.
4. Define the materials and processes to be used, namely, adhesives and cure cycle limitations.

Most repairs to aircraft composites involve damaged and possibly wet honeycomb in addition to the composite. The first priority is to prepare the composite for repair, which may involve drilling holes for a bolt-on patch or reducing the moisture to a level that will allow heating of the composite and the adhesive for cure.

Repair can involve the simple injection of a wet resin through drilled holes into the delaminations, then adding blind fasteners to defer the onset of local buckling. Alternatively, it can involve the use of wet or dry prepregs in conjunction with adhesives to form a plug, or a flush bonded-on patch. The available techniques for the repair of solid and honeycomb laminates are summarized in Table 4-1.

Bolted repairs are the most common for battle-damage repairs because of their reliability. Alternative bonding methods require more rigorous control of surface treatments, storage of heat-sensitive bonding materials, and special equipment and/or expertise (Ref 10).

Metal Matrix Composites

The trend in MMCs is toward fibers with improved performance and higher-temperature operating windows. The matrix trends are toward higher-temperature metals and lower-density metals. The trend in material systems is toward increased fracture toughness and durability. The trend in fabrication and inspection of MMCs is toward lower-cost methods, especially automation.

Table 4-1 Techniques for repairing solid and honeycomb composite laminates

Method	Advantages	Disadvantages	Ease of repair	Structural integrity
Bolted patch	No surface treatment; no refrigeration, heating blankets, or vacuum bags required	Bolt holes weaken structure; bolts can pull out	Fast	Low
Bonded patch	Flat or curved surface; field repair	Not suitable for high temperatures or critical parts	Fast, but depends on cure cycle of adhesive	Low to medium
Flush aerodynamic	Restores full design strength; high-temperature capability	Time-consuming; usually limited to depot; requires refrigeration	Time-consuming	High
Resin injection	Quick; may be combined with external patch	May cause plies to separate further	Fast	Low
Honeycomb; fill in with body filler	Fast; restores aerodynamic shape	Limited to minimal damage	Fast	Low
Honeycomb; remove damage; replace with synthetic foam	Restores aerodynamic shape and full compressive strength	Some loss of impact strength, gain in weight	Relatively quick	High
Honeycomb; remove damage; replace with another piece of honeycomb	Restores full strength with nominal weight gain	Time-consuming; requires spare honeycomb	More difficult	High

Large vertical tail boxes have been fabricated from the largest sheets of MMCs made in the United States. Large SiC-Al sheets, each 2.3 mm thick and measuring over 1780 by 590 mm, were fabricated in an Air Force Flight Dynamics Laboratory demonstration program. Assembly tooling for the project was similar to tooling used for metal-bodied airplanes. The only difference in the handling of the composite materials was in machine cutting, with feeds and speeds having to be adjusted somewhat. The sheets were handled through the normal production process to transfer as much technology as possible to the manufacturing floor. Fixtures and fittings were those normally used on metal-skinned airplanes. Skins for the tail boxes were cut with water jet cutters, and once the skins were cut to size they were fastened to MMC spars with standard steel fasteners, used to reduce costs. Four vertical tails were fabricated. Two had skins made of SiC whiskers mixed with aluminum, with spars made of SiC fiber mixed with aluminum. The other two tails had skins and spars made with SiC fiber mixed with aluminum. These were 2124 Al/SiC/15w and 6061 Al/SiC/42f composites.

Current classes of aerospace structures can benefit from the use of MMC. Weight savings is the primary benefit, which translates into various performance improvements. A further benefit of MMCs to aerospace structures is the dimensional stability that can be designed in. Because of higher operating temperature capability, there are numerous hot applications for internal engine components or hot airframe structures. Because the matrix material is not based on an organic compound, the outgassing problems inherent in high vacuum (space) applications are minimized (Fig. 4-5).

The MMC technology gaps are typical for a material in the early stages of development. Gaps exist in the engineering design and analysis databases for areas such as thermal fatigue, fracture, and joining.

Ceramic Matrix Composites

The trend in CMCs is toward fibers and whiskers with improved performance and higher-temperature operating windows, high-temperature matrices with improved chemistry, and material systems with increased fracture toughness and durability. The trend in fabrication and inspection of CMC materials and resulting structures is toward lower-cost methods and automation (Fig. 4-6).

CMCs have the same technology gaps that exist for MMCs, plus certain gaps in the materials engineering area, such as poor reinforcement-matrix interactions and lack of reliable forming technology.

Carbon-Carbon Composites

The trends in CCCs are improved oxidation resistance at ultrahigh temperatures and improved in-plane mechanical properties via higher strength. Improving the thermal stability of fibers is a trend, but the main emphasis is on improvement of matrix-dominated properties. The trend in fabrication and inspection of CCCs and resulting structures is toward lower-cost methods and automation (Fig. 4-7).

CCCs have the same technology gaps that exist for MMCs, plus certain gaps in the materials engineering area, such as low matrix strength and the lack of oxidation-resistant coatings.

Composite Materials in Aerospace Applications

Composite materials have demonstrated the potential for dramatic improvements in the performance of aerospace structures (Ref 1, 11). While a great deal has been accomplished, especially with regard to PMCs, it is important that a concentrated technical effort be pursued on a national scale to ensure that the competitive

Parameter	Trend
Reinforcements	Higher-strength fibers, SIC particulates/whiskers, ceramic fibers, B_4C fibers
Matrices	Higher-temperature matrix (copper, superalloys, etc.)
Processing	Low-cost castings, hot isostatic pressing, streamlined etrusions, liquid metal infiltration, diffusion bond
Manufacturing	Shear spinning, low-cost methods, fabrication in space, high-speed machining/forming/joining
Quality assurance	NDI of fiber-reinforced metal matrix composites, CAI/CII, AI process controls
Repair	Repair of fiber-reinforced metal matrix compsites, repair in space, joints
Performance Demonstrations	Space structure, HV structure, highly stable guidance control mounts
Design/Analysis	Design handbooks, durability analysis, computer integrated engineering

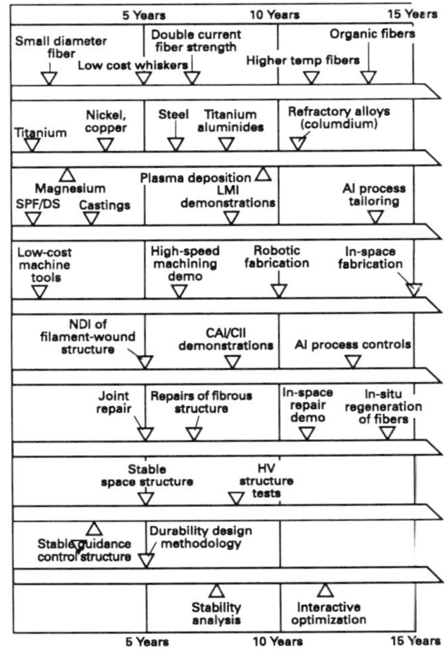

Fig. 4-5 Forecasts for metal matrix composites. AI, artificial intelligence. CAI, computer-assisted inspection.. CII, computer-integrated inspection. HV, hypervelocity vehicle. LML, liquid metal infiltration. NDI, nondestructive inspection. SPF/DB, superplastic forming/diffusion bonding. Source: Ref 1

Parameter	Trend
Reinforcements	Higher-strength fibers, SIC fibers and whiskers
Matrices	Improved-chemistry high-yield polymers, low-cost polymers, high-temperature matrices
Processing	Low-cost castings, hot isostatic pressing, liquid infiltration, chemical vapor infiltration, self-propagating high-temperature synthesis, microwave sintering
Manufacturing	Laser machining, low-cost processing, near-net-shape processing
Quality assurance	NDI of CMC, to < 50 μm flaw CAI/CII, AI process controls
Repair	Repair of CMC, repair in space, joints
Performance Demonstrations	HV structure, propulsion system components, space structures
Design/Analysis	Reliability analysis, computer integrated engineering

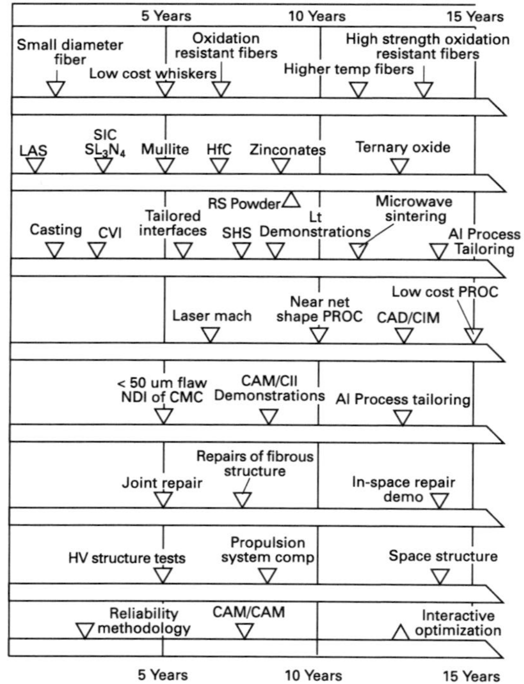

Fig. 4-6 Forecasts for ceramic matrix composites. AL, artificial intelligence. CAD, computer-assisted design. CAL, computer assisted inspection. CAM, computer-assisted manufacturing. CII, computer-integrated inspection. CIM, computer-integrated manufacturing. CMC, ceramic matrix composites. CVI, chemical vapor infiltration. HV, hypervelocity vehicle. LAS, lithium aluminosilicate. NDI, nondestructive inspection. SHS, self-propagating high-temperature synthesis. Source: Ref 1

Parameter	Trend
Reinforcements	Higher-strength fibers, heat-resistant and matrix-compatible fibers
Matrix Improvements	Higher-strength matrices, OIS matrix for cyclical environments
Processing	Low-cost processing and bonding techniques
Manufacturing	Low-cost methods, CAM/CIM, scale up to large parts
Quality assurance	CAI/CII, Integral oxidation indicators, internal sensors, improved computed tomography
Repair	Repair of CCC structure, joints
Performance Demonstrations	Airframes, space structure, propulsion systems
Coatings	Improved quality, uniformity, and high-temperature stability; scale up to large parts
Fastening/Joining	High-temperature high temp fasteners, bonding, bearing

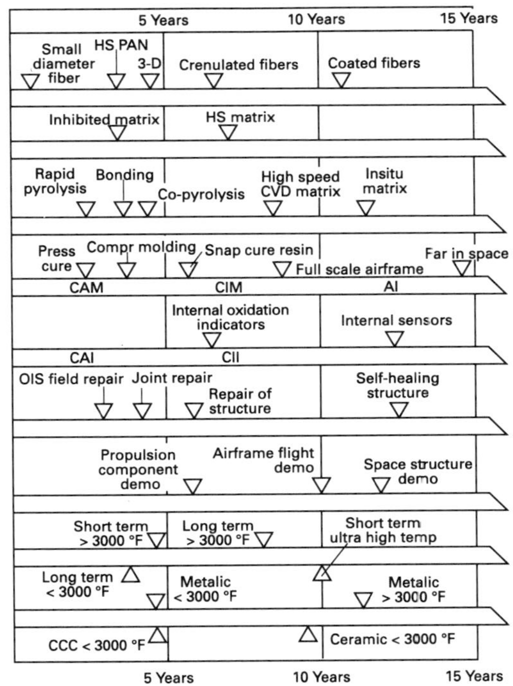

Fig. 4-7 Forecasts for carbon-carbon composites. Al, artificial intelligence. CAI, computer-assisted inspection. CAM, computer-assisted manufacturing. CCC, carbon-carbon composites. CIL, computer-integrated inspection. CIM, computer-integrated manufacturing. CVD, chemical vapor deposition. HS, high-strength. OIS, oxidation inhibition system. Source: Ref 1

position of U.S. industry remains strong. Currently, a number of other nations are making dramatic moves in the area of composite materials that ultimately will erode the U.S. position. For future systems, the materials and methods outlined in Fig. 4-8 must be developed. There is a vital need for a combined effort between U.S. academia, industry, and government to ensure that the necessary resources be available to keep the U.S. in the forefront of materials development and processes. There is a strong requirement for skilled labor, in addition to the need for equipment and financial resources.

Mechanical Fasteners for Aerospace Applications. In order to meet the manufacturing requirements for the intermediate sections of the B-2 stealth bomber, a computer-controlled robotic system for automated drilling and fastening of composite-metallic aerostructures was installed by the LTV Aerospace and Defense System Group. The automated five-axis robotic system performs the drilling and fastening operations of large, highly-contoured aircraft parts that combine composite materials with titanium or aluminum substructures. The aircraft sections are fixed while the computer-controlled robot moves over the assembly, drilling holes and installing fasteners in one pass in conjunction with specially-designed coolant-feeding drill/countersink tools. To date, the robotic cell has installed more

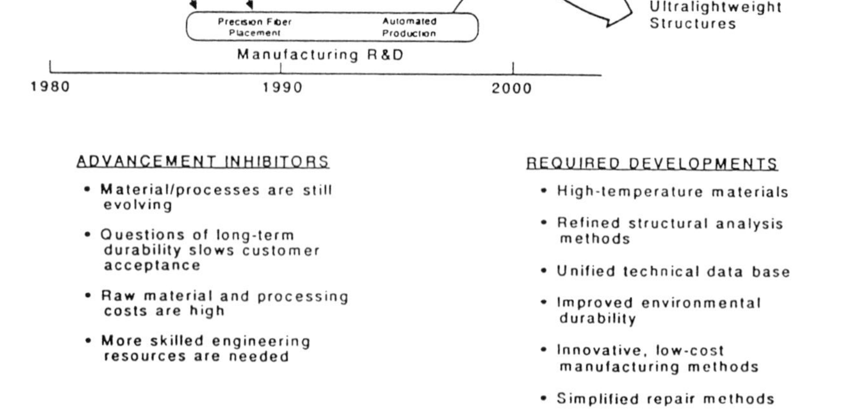

Fig. 4-8 Composite materials roadmap for aerospace applications. Source: Ref 1

than 50,000 fasteners in B-2 parts, improving the quality of fastener operations, decreasing production time, and reducing manual labor.

In another advancement in materials development in aircraft, the Boeing Co. is using a PEEK rivet to secure the liners of aircraft bays. The main reason for the selection of PEEK was that it features extraordinary flame retardance and withstands exposure to high temperatures. Although PEEK is not a new material, fasteners are a new application. This challenging application is expensive, but Boeing specifications have been satisfied.

The AV-8B Harrier wing assembly is shown in Fig. 4-9. Most spars and ribs composing the structure are made of a carbon-epoxy composite material. The remaining members are aluminum or titanium machined members. Most of the carbon-epoxy spars and ribs incorporate a sinusoidal-shaped web. The upper and lower carbon-epoxy skins, one-piece and highly contoured, are fastened to the substructure with approximately 6700 flush head screws and Hi-Lok fasteners (Ref 2).

To handle this fastening problem (Ref 12), an automated drilling system (ADS) composed of two numerically controlled, six-axis, gantry-type machines were developed to produce close-tolerance, outer-mold-line fastener holes in highly contoured parts. Numerous studies concluded that drilling holes with the ADS and installing fasteners to provide the clamping force was the most practical solution.

Installing a fastener in every tenth hole provided sufficient clamping action to install 600 holes immediately after each one is drilled, without causing excessive

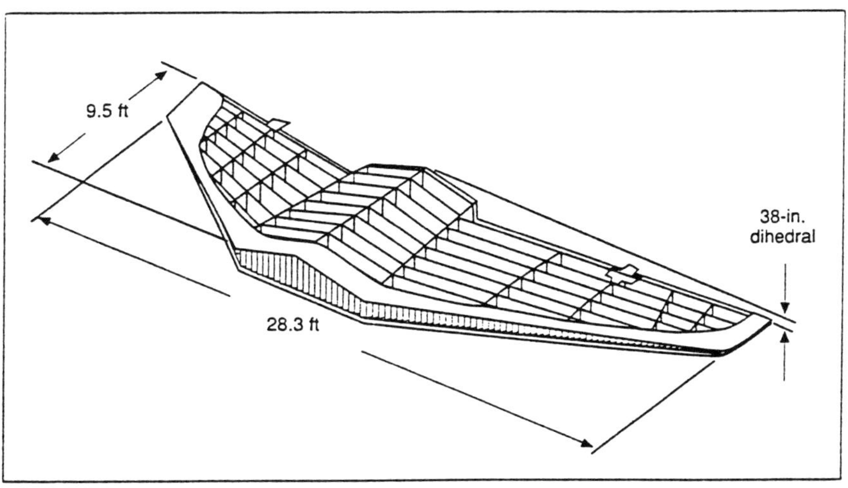

Fig. 4-9 AV-8B wing assembly. 25.7 mm = 1 in. Source: Ref 12

Table 4-2 Fastener characteristics for AV-8B Harrier wing assembly

Fastener characteristic	Contact area	Installed profile	Clamp force range	Clamp force control	Shank clearance	Diameter range	Grip range	Available reuses	Cost per fastener	Number of components	Access requirements	Removal efficiency	Installation efficiency	Proven technology
Fastener type														
Low-profile, single-side	U	...	...	...	...	...	...	...	...	...	...	...	...	...
Low-profile, single-side and washer	E	A	E	A	A	A	A	A	A	A	A	E	A	E
Screw, washer and nut	E	E	E	E	E	E	E	E	E	A	A	A	A	E
Standard-profile, single-side and washer	...	U	...	...	...	...	...	...	...	...	...	...	...	...
Pull rivet	...	...	...	...	...	...	...	...	...	...	...	U	...	...
Blind bolt	...	...	...	...	...	...	...	...	...	...	...	U	...	...
McAir design, single-side	A	A	E	A	A	A	A	A	P	E	E	E	E	P
Spring-loaded, single-side	...	...	U	...	...	...	...	...	...	...	...	...	...	...

E = excellent, A = acceptable, P = poor, U = unacceptable

reduction in rate capability or increasing the cost. The best fastener characteristics and their effect on the application are described in Table 4-2. The fastener installation system is illustrated in Fig. 4-10.

Fasteners come in all forms, including hybrids. The Aerospace Division of Standard Pressed Steel Technologies (Ref 13) faced a unique fastening problem with an advanced composite component for an aircraft fuselage. The problem was to form a blind head using a blind fastener in a 25.4 or 50.8 mm honeycomb joint between two composite parts. The fastener consistently pulled through the hole on the blind side and damaged the honeycomb material. The solution was a specially designed fastener made of titanium, two types of stainless steel, and a proprietary high-strength, corrosion-resistant alloy. The part didn't use the surface of the joint to form the blind side, because it had a coiled split washer that pulled up an angled ramp on the body of the fastener, simultaneously forming a sleeve in the fastener. This was accomplished without contacting the joint.

Virtually all fasteners require holes of some kind, but drilling a hole in some high-performance composites can delaminate the structure. Grommets should be

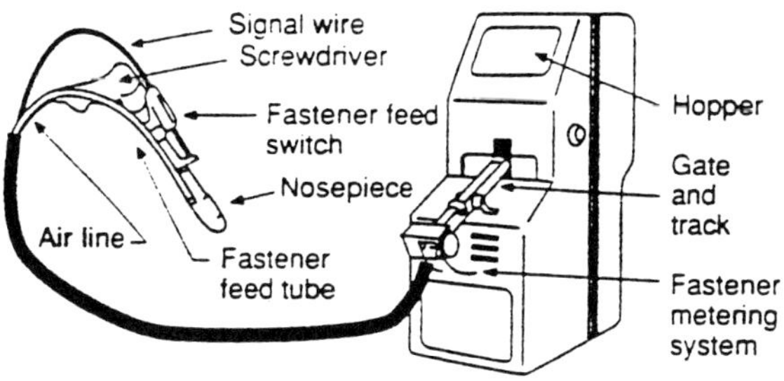

Fig. 4-10 Fastener installation subsystem used for AV-8B Harrier wing assembly. Source: Ref 12

used for hole support, especially with carbon and graphite composites if the hole isn't molded in when the composite is formed. A recently introduced threaded metal stud can be ultrasonically welded into thermoplastics, saving the cost and cycle time of making preformed holes.

Composite materials offer tremendous potential for meeting the performance goals of future systems. The need exists to develop all four categories of composites shown in Fig. 4-8. In the case of PMCs, emphasis should be placed on more innovative design and analysis, improved producibility, greater durability, and overall enhancement of the affordability of the finished component. For the three other categories of composites, emphasis must be placed on the evolution of material systems that will realize their tremendous potential, especially for higher-temperature applications.

In all cases, there is a need to reduce overall material and processing costs for composite materials. Increasing the volume of materials used will significantly reduce these costs, but for this to occur, the long-term durability of these materials must be demonstrated in order to obtain increased customer acceptance. For composite material technology to develop, the number of trained engineering personnel must increase in all categories, especially materials science, engineering mechanisms, and manufacturing engineering. Finally, the development of a nationally accepted materials database is needed for full composite materials acceptance.

Adhesive bonding for aerospace applications requires consistent strength and long-term reliability. To achieve consistent strength, the bond line thickness must be controlled so that repeatable joints with similar performance characteristics are achieved (Ref 14). A number of decisions must be made before bonding to a composite is attempted.

The surface preparation technique differs for composite and metal. For composite, peel ply, manual abrasion, or co-cure should be used. For aluminum, phosphoric anodizing is the preferred airframe method. For titanium, several types of etches are available. For steel, the time between surface preparation and bonding is of concern.

The type of adhesive is also a factor. Film adhesive offers reproducible chemistry and allows early, frequent quality assurance checks (because resin is premixed by the manufacturer). However, it requires special storage and generally cannot accommodate varying bond line thicknesses. Paste adhesive offers longer shelf life and fewer or no changes in storage, which does accommodate varying bond line thicknesses. However, quality assurance must be performed after the structure is bonded.

The maximum cure temperature of the adhesive should be below the composite cure temperature unless they are co-cured. The cure temperature or the upper use temperature of the adhesive may dictate the maximum environmental exposure temperature of the component. There are several cure temperature ranges:

- Room temperature to 107 °C: Generally used for paste adhesives for noncritical structures.
- 107 to 140 °C: Used for nonaircraft critical structures. In general, this range cannot be used for aircraft because moisture absorption may lower the heat-deflection temperature below the environmental operating temperature.
- 177 °C and above: Used for aircraft structural bonding. A higher cure temperature may mean more strain discontinuity between adhesive and both mating surfaces (residual stresses in bond).

Joint Design. The most desirable method of load transfer through an adhesive-bonded joint is by shear. This means that the design must avoid peel, cleavage, and normal tensile stresses. Rivet-bonded construction avoids stresses other than shear stress. The direction of the fibers in the outer ply of the composite (against the adhesive) should not be 90° to the expected load path.

Because of the breadth of materials technology, it is difficult to synopsize all of the key trends for the next 15 years. What is apparent is that all classes of composite materials do offer significant potential for reduced structural weight and higher-temperature capability. The trends shown in Fig. 4-11 for weight, cost, and temperature capability should be considered qualitative measures of future improvements. For the less developed classes of advanced composites (i.e., CCCs and CMCs), these improvements would be much larger over the projected time frame.

Nanocomposites

The increasing requirements of all industries, but particularly aerospace, have led to the emergence of nanocrystalline materials for improved properties through fineness of structure. Nanostructures are single-phase or multiphase polycrystals with typical crystal size in at least one direction in the range of 1 nm to tens of nanometers (Ref 15). Because of their extremely fine grain size, nanocomposites have many more atoms located in grain boundary regions than conventional materials have (Fig. 4-12). These atoms impart novel characteristics such as very high diffusion rates along the "open" grain boundary regions and the possibility of alloying with normally insoluble constituents.

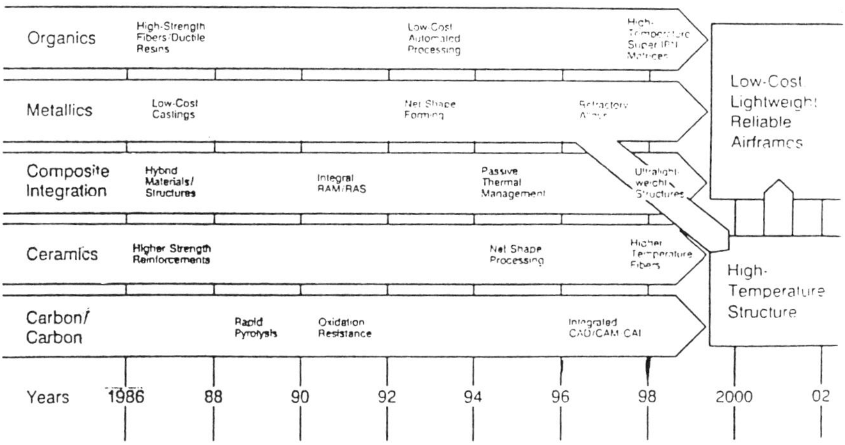

Fig. 4-11 Composite materials integration roadmap. CAD/CAM, computer-assisted design/computer-assisted manufacturing. CAI, computer-assisted inspection. IPN, interpenetrating polymer networks. RAM/RAS, radar-absorbing material/radar-absorbing structure. Source: Ref 1

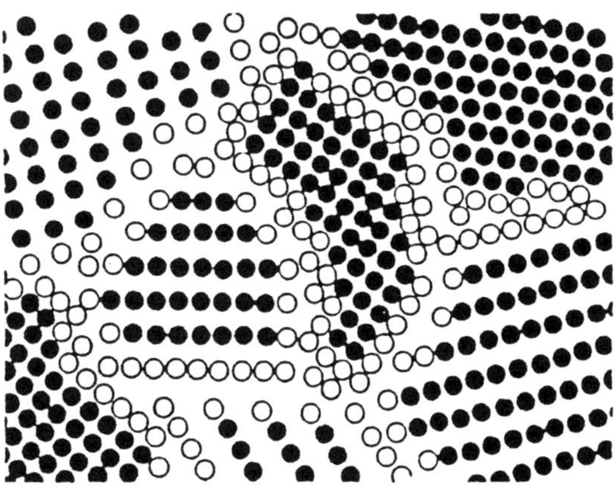

Fig. 4-12 Schematic drawing of the structure of an unrelaxed nanophase material. Filled circles, atoms associated with the individual crystal grains. Open circles, atoms constituting the interface network

The grain size, morphology, and texture can be varied to improve the mechanical and physical properties of conventional crystalline material. This is accomplished by modifying or controlling the thermal processes by which the nanocrystalline materials are prepared. Preparation is being done by a variety of processes, including rapid solidification from the liquid state, electron beam vapor deposition, ion beam, molecular beam epitaxy, reactive sputtering, mechanical alloying, sol-gel, electrochemical deposition, and gas condensation.

The use of nanocomposites in materials processing can lead to monophasic or multiphasic ceramics, glasses, or porous materials with tailored and improved properties. A variety of nanocomposites (e.g., sol-gel, intercalation, entrapment, electroceramic, and structural ceramic types) have properties that are superior to those of the monophasic or microcomposite alternatives, and they are forecasted to have a major impact in catalytic, sensor, optical, electroceramic, and structural ceramic materials (Ref 16). The investigation of nanostructures is in its infancy, but the potential of these materials and the many unexplored technological aspects that appear promising will intensify efforts in their development.

Powder metallurgy is an obvious fabrication technique for production of nanostructures, either using condensation from the vapor phase on a cold finger or by mechanical alloying methods, possibly even via an amorphous precursor. Table 4-3 gives examples of how behavior is modified in moving from conventional crystalline material to amorphous glassy product and finally to nanocrystalline structures. These nanostructures could well see application by the later 1990s (Ref 17, 18).

Chemical Vapor Deposition (CVD). Hirai et al. (Ref 19) are doing some very interesting work in which CVD is used to fabricate composites. The properties of

Table 4.3 Properties of nanocrystalline materials compared with those of their crystalline and glassy counterparts

Property	Material	Nanocrystal	Glass	Crystal
Thermal expansion (10^{-4}/K)	Cu	31	18	16
Specific heat at 295 K (J/gK)	Pd	0.37	...	0.24
Density (g/cm^3)	Fe	6	7.5	7.9
Elastic moduli				
Young's modulus (GPa)	Pd	88	...	123
Shear modulus (GPa)	Pd	32	...	43
Saturation magnetization at 4 K (emu/g)	Fe	130	215	222
Susceptibility (10^{-4} emu/Oeg)	Sb	20	−0.03	−1
Fracture stress (kg/mm^2)	Fe-1.8C	6720	...	560
Superconducting critical temp. (K)	Al	3.2	...	1.2
Activation energy for diffusion (eV)	Ag in Cu	0.51	...	2.0
	Cu in Cu	0.64	...	2.04
Debye temp. (K)	Fe	345	...	467

Source: Ref 46

composites are controlled by the geometric arrangement between the matrix and the dispersion. When the size of the dispersoid approaches nanometers, it is very difficult to fabricate composites by mixing powders followed by some type of consolidating process. Because of such problems, it is essential to prepare composites by using an approach in which phase separation occurs in situ. In-situ composite preparation with nanometer dispersoids can be achieved by using a gas (CVD, physical vapor deposition (PVD)), a liquid (sol-gel, coprecipitation), a molten alloy (solidification, eutectic reaction), or a solid (phase separation, grain boundary reactions). CVD is an excellent method to form nanoscale deposits as films on different substrates. Hirai et al. have prepared a number of nanoscale composites by using CVD; some typical ones are illustrated in Fig. 4-13.

In the case of a C-SiC composite having disks of SiC several nanometers in diameter, the carbon and SiC possessed an orientation relationship between each other, and it was reported that this composite had better oxidation resistance than carbon. When particles of carbon were dispersed in Si_3N_4, these particles formed a three-dimensional network, and the addition of 0.2 wt% C increased the electrical conductivity of the composite very substantially. Other examples have been presented (Ref 19) to show that nanocomposites can be prepared such that transparency is maintained at elevated temperatures, thermal conductivity is modified, and toughness is improved. Table 4-4 presents some nanocomposites prepared by CVD and the properties altered by using this approach (Ref 20). Also under study are nanocomposites of nonoxide ceramics prepared by CVD.

Sintering. Scientists from Osaka University (Ref 21) have developed intergranular and intragranular ceramic nanocomposites by normal sintering. In their work, alumina, mullite, magnesia, yttria, and silicon nitride matrix materials with 3 to 5 vol% reinforcements of nanosize silicon nitride, silicon carbide, and titanium diboride had higher room-temperature toughness and strength than monolithic ceramic materials.

Hot Pressing. Nanocomposites have also been fabricated by hot pressing a mixture of SiC and Si_3N_4 fine powders. SiC particles have been dispersed uni-

Table 4-4 Some examples of nanocomposites prepared by chemical vapor deposition

Matrix	Dispersoid	Dispersoid shape	Dispersoid size (nm)	Property altered
C	SiC	Disk	3-30	Oxidation resistance
Si_3N_4	BN	Lamellar	≈300	Transparency
Si_3N_4	C	Spherical	≈300	Electrical conductivity
SiC	TiC	Spherical, fiber	30-300	Toughness
SiC	C	Spherical, thin layers	30-300	Thermal conductivity
Fe, Ni	FeB_x, NiP_x	Spherical	30-300	Strength

Source: Ref 20

formly in Si_3N_4 grains and condensed at grain boundaries of Si_3N_4. This is very important in composite materials in particular, because the interface of different phases is an important factor in their mechanical properties (Ref 22).

Other issues in the hot pressing of nanocomposites are being studied:

- To create a satisfactory density for Al_2O_3/5 vol% TiC nanocomposites, a hot-pressing temperature of 1600 °C was required. This resulted in the fracture strength of Al_2O_3/5 vol% TiC nanocomposite of 1050 MPa for a

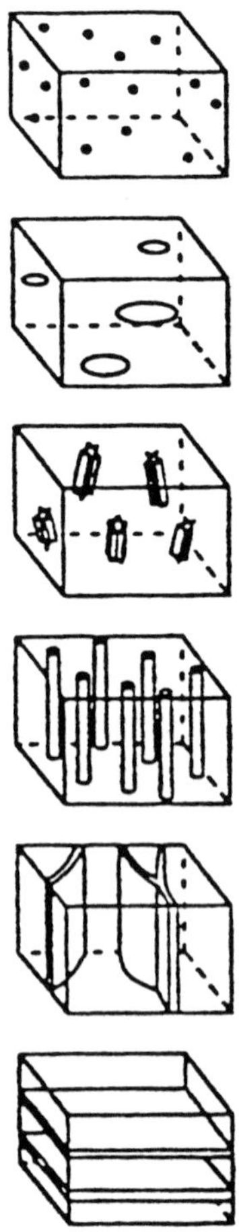

(a) Spherical particle dispersion (C-B_4C). C-TIC, C-ZrC, SiC-C, Si_3N_4-C, Si_3N_4-TiN, Si_3N_4-AlN, BN-TiN, BN-Si_3N_4).

(b) Disk-type particle dispersion (C-SiC).

(c) Rod-shaped particle dispersion (BN-BN).

(d) Fiber dispersion (Si_3N4-Tin, SiC-TiC).

(e) Thin layer dispersion (Ti_3SiC_2-TiC).

(f) Lamella (Si_3N_4-BN).

Fig. 4-13 Nanostructures of nanocomposites produced by chemical vapor deposition

sintered body at 1600 °C. The composite material satisfactorily reinforced showed a high strength of more than 1200 MPa (Ref 22).

- Nanocomposites have been fabricated by hot pressing the mixed powders of ultrafine Al_2O_3 and SiC. Transmission electron microscopy of these Al_2O_3/SiC nanocomposites showed that SiC particles less than 200 nm were uniformly dispersed within the Al_2O_3 matrix grain, and dislocation networks were seen within Al_2O_3 matrix grains (Ref 22).
- The effects of TiC and ZrO_2 dispersions on the fracture toughness strength and hardness of Al_2O_3/TiC/ZrO_2 composites are being studied (Ref 22).

Molecular Machines. Within the next 15 to 20 years, molecular machines for materials processing will be developed. This new application of molecular nanotechnology will occur for the bulk processing of materials and the fabrication of custom products. Molecular nanotechnology relies on the use of a large number of small manufacturing systems working in parallel and using locally available materials. Built to atomic specification, the products exhibit order-of-magnitude improvements in mechanical properties and are of high quality and low cost.

Since the introduction of the concept in 1981 by K.E. Drexler (Ref 23, 24), who detailed the design, performance, and development path of these systems, major advances toward the development of molecular nanotechnology have been recorded. For example, scanning probe microscopes can be used to manipulate individual atoms to subangstrom precision, molecular biologists can design and synthesize proteins from scratch, chemists can design and create uncomplicated molecules to trap ions and molecules, and research on molecular electronic devices has already resulted in a working prototype. In addition, the wide availability of scientific workstations and molecular modeling software is significantly reducing the amount of time required to electronically construct and test new designs (Fig. 4-14, 4-15). Drexler contends that nanosystem components would realize capabilities of tensile strength $\geq 5 \times 10^{10}$ GPa (Ref 23).

Functionally Gradient Materials

Functionally gradient materials (FGMs) are composites in which the proportions of the constituent phases are varied in an article of interest made from the composite, in order to tailor the properties locally to improve the performance of the article. A typical FGM would provide the requisite heat resistance to a surface in contact with gas that is heated to several thousands of degrees, by:

- Arranging ceramics on the surface to effect heat conductivity
- Providing the necessary mechanical strength by arranging a metallic material

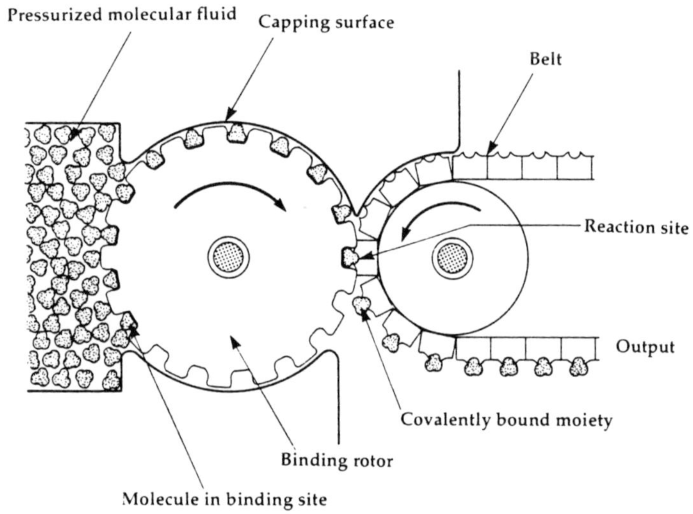

Fig. 4-14 Schematic drawing of molecular machine. Feedstock molecules are transferred from a pipe onto a conveyor belt for transport to the assembly site, where they will be bound to the finished structure. Source: Ref 24

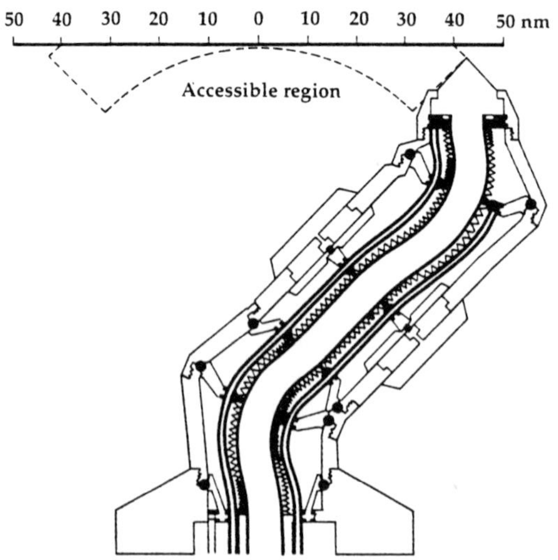

Fig. 4-15 Schematic drawing of molecular machine. Manipulator arm positions molecules to form an atomically precise structure. Source: Ref 24

- Positively relaxing thermal stresses by synthesizing materials so that their composition, structure, and the pattern of holes between the two are optimized.

Surface temperatures as high as 2000 °C and temperature gradients on the order of hundreds of degrees per millimeter are envisioned for these materials. Methods of FGM evaluation include laser thermal impact tests, xenon arc beam and burner heating, and plasma flame heating tests, which also serve as high-temperature oxidizing tests.

FGM Fabrication. For space plane applications, materials must be able to withstand temperatures of 1800 °C and temperature gradients of 1300 °C. Therefore, the materials must have suitable heat resistance and oxidation resistance in the surface layer, mechanical toughness in the cool surface, and an effective thermal-stress relaxation throughout the material. Six main approaches have been identified for producing such FGMs: powder metallurgy, plasma spraying, vapor

Table 4-5 Current research in functionally graded materials

Organization	Composition	Process	Comments
Tohoku Univ. Sendai, Japan	Partially stabilized zirconia/stainless steel	P/M	(a)
Nippon Oil Co. Yokohama, Japan	SiC-coated CCC	CVD	Good high-temperature fatigue, complex shapes
National Res. Institute for Metals Tokyo, Japan	ZrO_2-Ni alloy	Plasma spraying using twin torches	Thin layers(b)
Institute of Scientific & Industrial Res. Osaka Univ., Osaka, Japan	TiC-Ni	Gas-pressure CS	(b)(c)
Government Industrial Res. Institute Tohoku, Japan	TiB_2-Cu	CS combined with hydropressing	Room-temperature short processing times
Hokkaido Univ. Sapporo, Japan	Ni-NiO, Ni-MgO, Si_3N_4, Al-AlN	P/M	(c) for joining applications
Alfred Univ. Alfred, N.Y.	TiC-NiAl	CS	Joining applications, thermal stress function of sample geometry
Ohio State Univ. Columbus, Ohio	NiAl-Al_2O_3	Sedimentation in organic solvents	Additives reunited to prevent segregation, drying shrinkage causes cracking
University of Idaho Moscow, ID	Ti-B-Cu	CS	Intermetallic phases reduce toughness

P/M, powder metallurgy; CS, combustion synthesis, CVD, chemical vapor deposition. (a) Thermal stress reduction dependent on width and composition profile, sample geometry, and dimension. (b) Near-net shapes. (c) Lower-temperature processing

deposition, self-propagating high-temperature synthesis, thin film lamination, and sintering synthesis. Table 4-5 summarizes the current research.

Powder metallurgy uses blended metal and ceramic powders that are dispersed in a suitable liquid vehicle. Successive layers of the slurries, with increasing concentration of the ceramic constituent, are sprayed onto a metallic base. The composition of the metal powder corresponds to that of the base metal or alloy. After a sufficient number of sprayed layers, the composition of the slurry is 100% ceramic. The sophisticated control of this process and the grading that can be achieved are shown schematically in Fig. 4-16 and 4-17.

Various materials systems have been investigated to date: 304 stainless steel/ZrO_2, 304 stainless steel/Si_3N_4, Ni/ZrO_2, Mo/ZrO_2, and Mo/Si_3N_4. The powders of partially stabilized zirconia/stainless steel (PSZ/SS) were given con-

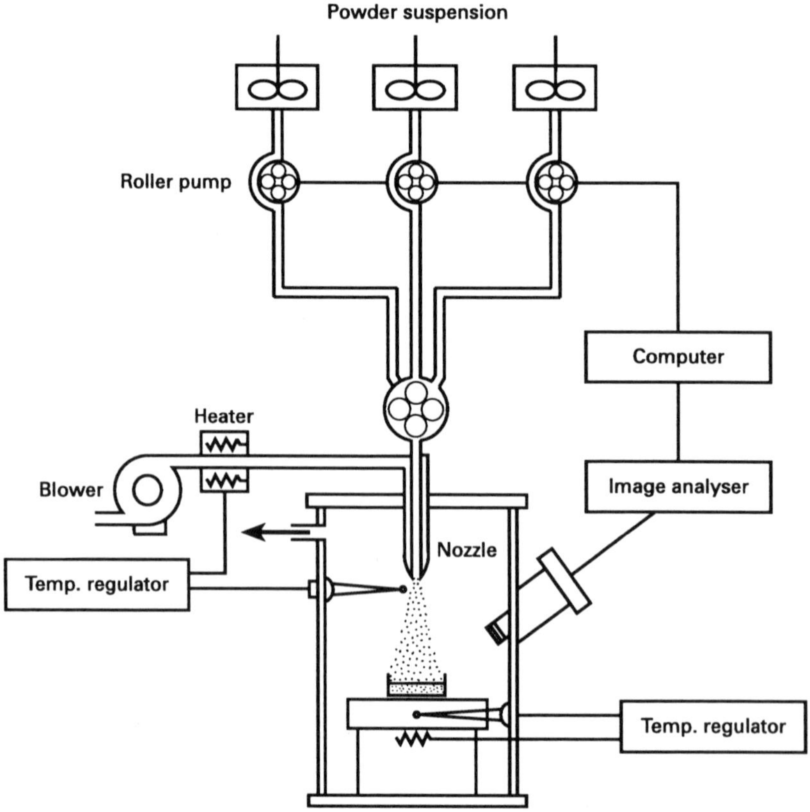

Fig. 4-16 Schematic drawing of powder spray stacking apparatus. Source: Ref 25-27

tinuously changing compositions by stacking layers on layers, each layer altered accordingly to predesigned composition profiles. The layered powders were pressed to shape and sintered. The rapid reactions that occurred in the combustion synthesis process proved useful in this FGM work.

Fig. 4-17 Schematic drawing of microstructural transition with compositional variation in a functionally gradient material produced by powder metallurgy. Source: Ref 25-27

Metallurgists in Japan (Ref 28) have produced titanium diboride/copper and titanium carbide/nickel systems that might eventually prove useful for aerospace structures such as rocket thrust chambers. Although aerospace and nuclear uses prompted the original interest, the versatility and tailorability of FGMs have made them attractive for optical devices, sensors, and electronic components.

Joint research at Hokkaido University and Nisshin Steel Co. involves using a powder metallurgy method to prepare FGMs of Ni/MgO, Ni/NiO, Ni/Si_3N_4, and Al/AlN. A mixture of metal and ceramic powders with compositional gradient is cold pressed at 20 to 32 MPa.

In order to design a satisfactory FGM, it is necessary to have a predictive model or else have measurements of the mechanical strength, elastic, and thermal properties as a function of the composition of the layers. To guide specifications for thicknesses, compositions, and powder characteristics, computer programs that incorporate elements of design and materials knowledge bases have been developed.

The powder approach appears to be relatively straightforward, and it offers a wide choice of primary constituents, as well as good microstructural control. Depending on the shape of the part or component, it may be possible to further enhance the quality of the composite by means of pressure-assisted sintering.

Plasma Spraying. A number of plasma spraying methods have been developed to prepare PSZ-based systems. The National Institute for Metals uses twin torches in air for making graded coatings of PSZ-Ni alloys. Nippon Steel Corp., on the other hand, uses a single gun with four ports at low pressures for making PSZ-NiCr coatings. The ceramic and metal powders are supplied to the plasma arc simultaneously. Plasma spraying methods are expected to have the capability to produce large-scale, complex structures and the capability for mass production.

Vapor Deposition. By gradually modifying the mix of the vapor species from which the separate constituents of an FGM are deposited, it is possible to produce a gradual shift in the composition of the deposit. The microstructure of CVD-produced materials is often quite different from that of materials produced by powder metallurgy. It is not unusual for a CVD deposit to have crystal orientations or shapes that are different in the deposition direction from those in the direction parallel to the deposition surface.

For the CVD approach to be applicable to FGMs, the separate constituents must deposit under similar sets of conditions (e.g., temperature, pressure, and composition of the reacting gas). These conditions also depend on the choice of the starting materials. Once conditions for the codeposition of two separate materials are established, not only FGMs but also nanocomposites can be prepared.

An FGM system that has received considerable attention is C-SiC (Ref 22, 29). In this case the purpose of the SiC outer layer is to provide protection against oxidation of a carbon structure. Such a coating could allow carbon to be used as a combustor or allow similar applications in which it is exposed to an aggressive environment. Codeposition results when a mixture of $SiCl_4$ and C_3H_8 ($CH_3 \cdot CH_2 \cdot CH_3$ = propane = C_3H_8) is reacted with hydrogen gas at 1 atm total gas pressure at 1500 °C. The gradual transition from carbon to SiC deposition is achieved by the gradual shift of the $C_3H_8/SiCl_4$ ratio from infinity to zero. Pure C_3H_8 plus hydrogen gas leads to a smooth carbon coating that adheres to the bulk carbon substrate.

In real-condition tests, an FGM must be actively cooled to generate a high-temperature gradient in its depth direction. To ensure proper cooling between tests, bonding of the FGM and an actively cooled panel is very important. The CCC FGM cooling panel used by Yamaoka et al. in an experiment (Ref 30) consisted of layers of C-SiCl. Oxygen-free copper was used as the material for the cooling panel because of its high thermal conductivity. A silver braze filler material produced excellent bonding results vis-à-vis the CCC when brazing was conducted at 900 °C for 60 min at 0.026 Pa. As a result of these efforts, a method for brazing a CCC plate to an oxygen-free copper cooling panel was established.

Hirai et al. (Ref 31, 32) are preparing FGMs by using vapor deposition processes. Compositionally graded films of Ti-TiC on carbon steel, and Ti-TiN on titanium and copper have been fabricated. An FGM film graded from SiC to carbon has also been fabricated on graphite by using $SiCl_4$-CH_4-H_2 at temperatures of 1399.85 to 1499.85 °C. This FGM film did not crack under cyclic high-temperature heat flow conditions, whereas SiC on graphite with no FGM did crack. Figure 4-18 shows how this SiC-C FGM is proposed to be used to protect a CCC.

Other FGMs being developed are Y_2O_3-stabilized ZrO_2 and Ni-Co-Cr-Al-Y on Hastelloy C and on stainless steel substrates. The FGMs are about 600 μm thick and are graded from 100% stabilized ZrO_2 to 100% Ni-Co-Cr-Al-Y. The specimens have been subjected to a temperature differential of 500 °C along with one-dimen-

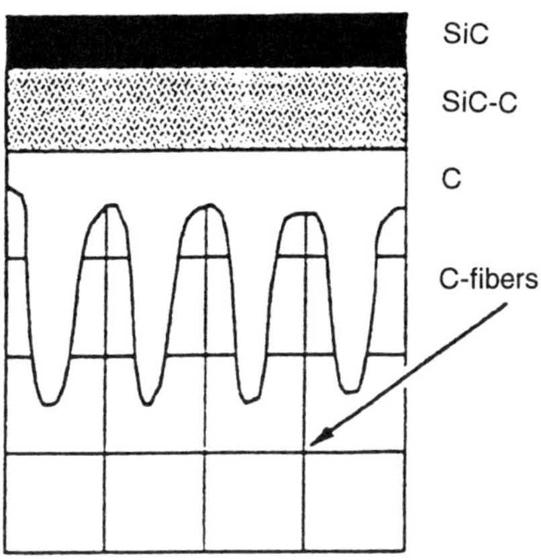

Fig. 4-18 Schematic drawing of a SiC/C functionally gradient material on a carbon-carbon composite. Source: Ref 31

sional tensile loads. It has been determined that these FGM coatings have good bonding properties compared with those of conventional thermal barrier coatings.

Other investigators are developing FGMs on TiAl. The surface of the FGM for high temperature is SiC. The composition is graded from 100% SiC to mixtures of SiC and TiC to 100% TiC on TiAl.

At the University of Idaho (Ref 28), researchers produced a Ti-B-Cu composite in which the titanium exhibited a graded percentage change from 70 to 40%. The copper portion exhibited a 10 to 40% change throughout the material.

Self-propagating high-temperature synthesis (SHS) is also known as combustion synthesis or reactive sintering. At Osaka University, SHS was combined with HIP to prepare TiC-Ni. The pressed body of reactant powders was vacuum sealed into a glass capsule and placed into a crucible filled with a combustion agent. The reaction was initiated by the heat released from the combustion agent under argon pressure below 100 MPa. Flexural strengths at room temperature and 500 °C ranged from 1100 to 1340 MPa and 630 to 920 MPa, respectively, depending on composition.

Gravity-Based Sedimentation in Organic Solvents. FGMs of NiAl-Al_2O_3 have been fabricated by sedimentation in organic solvents, followed by densification by hot pressing. The advantages of this method include extensive control of the gradient composition and control of second-phase additions. However, segregation of ceramic and metal powders can occur because of differences in particle size and density. This problem can be solved by choosing appropriate additives.

Sample Application for FGMs. A fundamental study of an FGM brazed to metal plate was carried out to develop the fabrication of an actively cooled panel with an FGM (Ref 33). The FGM was designed to reduce thermal stress in service where the top surface was heated heavily and the bottom surface was cooled with coolant, resulting in a large temperature difference between the surfaces. The FGM was constituted of ZrO_2 and nickel, and the composition was varied gradiently from the top surface to the bottom surface. Thus, a layer in the FGM was different in CTE from any other layer, and the shape of the FGM inherently changed as its temperature varied uniformly, as in the case of brazing.

The FGM used in the research program was composed of six layers: ZrO_2 100%, ZrO_2 80%/Ni 20%, ZrO_2 60%/Ni 40%, ZrO_2 40%/Ni 60%, ZrO_2 20%/Ni 80%, and Ni 100%. It was created by the thin film lamination method. The outermost layers, ZrO_2 100% and Ni 100%, were formed by the plasma thermal spraying method and the electroforming method, respectively. The brazing test, which used BNi-2 (Ni-7Cr-3Fe-3B-4.5Si) filler metal, was effective in brazing the FGM plate to the metallic plate (Fig. 4-19). Liquid diffusion bonding brazing was used for isothermal solidification of the filler.

Functionally Gradient Ceramics. Moya et al. (Ref 34) examined FGCs (functionally gradient ceramics), materials having a continuously varying property (i.e., hardness, dielectric constant, bioactivity, etc.) from one surface to the other. A sequential slip casting technique was studied as a new way to produce complexly

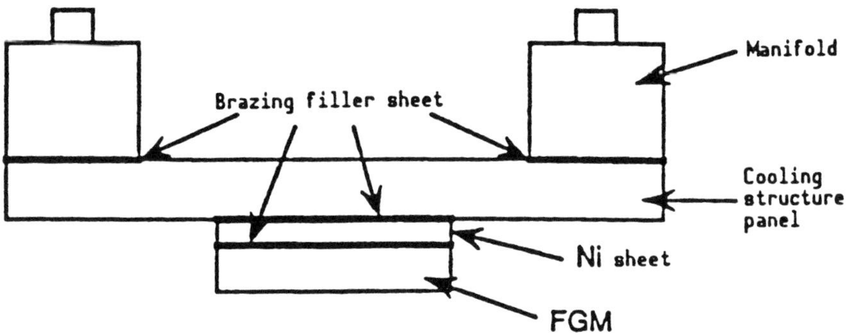

Fig. 4-19 Two-stage brazing in fabrication of a cooling panel from functionally gradient material. Source: Ref 33

shaped FGCs with close to theoretical density. Al_2O_3 and yttria/ZrO_2 were selected because of their exceptionally high mechanical performance and consequently their high potential structural applications. The results show that with sequential slip casting there is potential to develop a new family of FGCs with complex shapes and homogeneous composition of the different layers. The specifically tailored microarchitecture of these materials would make it possible to face technological challenges that would be impossible to face with monolithic materials.

Intermetallic Matrix Composites

IMCs are currently of extreme interest for future high-temperature, high-efficiency, high-performance gas-turbine engines for civil and military applications. Some of the composites under consideration for aerospace applications are:

- *Matrices:* FeAl, Ti_3Al, TiAl, Ni_3Al, NiAl, Nb_3Al, and $NbAl_3$
- *Fibers:* SiC, TiB_2, TiC, graphite, Al_2O_3
- *Advantages:* Light weight, stiffness, design flexibility, balanced matrix and fiber properties
- *Disadvantages:* CTE mismatch, chemical compatibility, fabrication, joining

The low density of IMCs can be used to good advantage, and if low-density, high-strength fibers are available, the low strength of IMCs becomes less of an issue. The matrix can be optimized for other properties, most importantly ductility, oxidation resistance, and density. The influence of the fiber on the strength of a composite has been discussed by McDanels and Stephens (Ref 15) for the use of SiC-reinforced composites, where the strength-to-density ratio was predicted to be

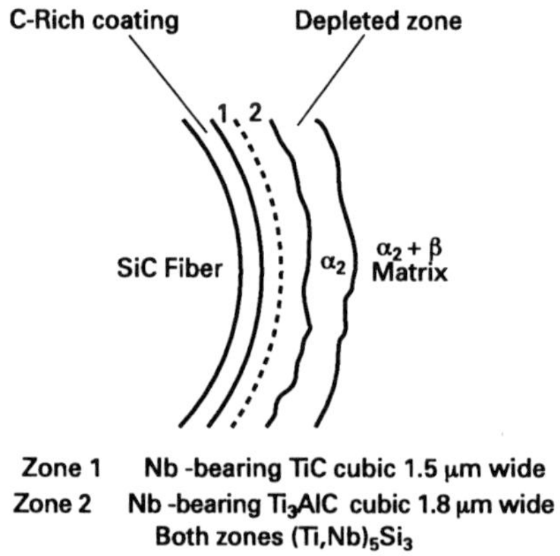

Fig. 4-20 Incompatibility of fiber and matrix in an intermetallic matrix composite

essentially independent of matrix strength. Experimental results for this concept are given in Ref 35. The tensile behavior achieved the predicted rule-of-mixture behavior, and on a density-corrected basis, the composite was shown to have tensile properties superior to those of wrought nickel-base alloys, cobalt-base alloys, and single-crystal superalloy NASAIR 100.

One of the major problems in the development of IMCs is the incompatibility between fiber and matrix, chemically and in terms of the CTE. An example of the extreme complexity of fiber-matrix interactions is shown schematically in Fig. 4-20 for an SCS-6 fiber in a Ti_3Al-Nb matrix. The SiC fiber has a two-layer carbon zone on the surface that further contributes to the chemical reactions in this system. To overcome the fiber-matrix incompatibility problems, studies of the kinetics of reaction types and methodology have been undertaken to identify the most appropriate fibers for the intermetallic matrices (Ref 36).

IMC Fabrication

A floating zone single crystal fiber growth technique has been developed by NASA-Lewis Research Center (Ref 37). The efforts focused on optimizing the growth of Al_2O_3 fibers and the operation of the floating zone system. The first step in optimizing Al_2O_3 growth was to develop a process to efficiently make feed rods for fiber growth. From this experience, other fiber compositions, in particular the beryllides of niobium, zirconium, titanium, and tantalum, along with other intermetallic materials and some mixed compounds, were examined as promising materials for advanced reinforcements for IMCs. Another approach has been continuous CVD processing of TiB_2 fibers and investigation of the feasibility of depositing a compliant layer suitable for overcoming both thermal expansion mismatch and chemical reactivity.

Ti_3Al, or alpha, and TiAl, or gamma, are the two main titanium aluminide IMCs being pursued by NASA and the Air Force (Ref 38-41). Gamma is the more desirable of the two because it is lighter (more aluminum for each titanium atom) and has a higher projected use temperature. Gamma melts at 1482 °C and could be useful past 982 °C, while alpha may be usable past 704 °C. Other aluminides under development are nickel aluminides, which melt at 1649 °C, and niobium aluminide ($NbAl_3$), which melts at 1718 °C.

Carbon-coated SiC has successfully reinforced alpha, which has room-temperature ductility of 1.5 to 3%. Carbon coating decreases the propensity of titanium to mate with any available fiber when excited at high temperatures. Ti_3Al is ductile enough to adjust to the different CTE of SiC fiber. That is not the case with gamma. SiC reacts with TiAl when consolidated into a composite at 982 °C. SiC and TiAl also have different CTEs. Gamma ductility of only 0 to 0.5% offers no "give" when fibers and matrix expand and contract at different rates during thermal cycling, and the result is a cracked matrix. Research quantities of gamma with 3% room-temperature ductility have been produced, but high-temperature properties or scaling up results are unknown.

The other way to solve both chemical compatibility and thermal expansion problems is to change the fiber reinforcement. Two candidates are titanium diboride and oxide ceramics. CVD is used to grow titanium diboride, titanium silicide, and titanium carbide over resistively heated molybdenum fibers. Projected costs are roughly equivalent to those of SiC fibers. Titanium diboride is the most promising of the three, and preliminary tests show that it works well with TiAl. Good mechanical properties and good rule-of-mixture properties are being developed.

On the plus side, though, the CTE of titanium diboride more closely matches that of TiAl, and the material is less reactive at 982 °C than SiC. The fibers also conduct heat better than SiC for parts designed to draw heat away from the hottest sections of an engine or aerospace structure.

An in-situ reinforcement technique combined with an investment casting process developed at Martin Marietta and Howmet now produces economical near-net-shape parts in $TiAl/TiB_2$. In addition, composites containing TiB_2 particles introduced via exothermic dispersion (XD) processing offer substantial grain refinement and improved microstructural uniformity compared to that of unreinforced TiAl.

In rotating components, the low density of $TiAl/TiB_2$ reduces the strength requirements of the whole system. For example, reducing the weight of low-pressure jet engine turbine blades by 30 to 50% could allow weight reductions in the turbine disk, turbine shaft, and bearing support housings. Other examples of rotating applications include compressor vanes, rotating seals, diffusers, and turbocharger wheels.

The patented XD technology is based on sustained high-temperature synthesis. In this process, elemental boron, titanium, and aluminum powders are blended and compacted, then ignited. As the flame front passes through, titanium diboride and titanium aluminide form spontaneously. Because of in-situ development of reinforcement, the process eliminates oxide formation that could weaken the interface between the reinforcement and the matrix. The result is a master alloy with a high (20 to 75%) concentration of TiB_2, which can be added to a matrix material to achieve required levels of reinforcement (Ref 42).

In the Howmet adaptation of the technology, the XD master alloy is combined with other alloying elements to form an electrode, which is then triple-vacuum-arc remelted. The resulting ingots are well suited to investment casting because the reinforcements are stable in the liquid phase, retain their fine size through the investment casting process, and generate a fine, homogeneous structure. TiB_2 reinforcement levels for the ingots range from 0.8 to 7 vol%.

Al_2O_3 is the most familiar ceramic oxide fiber. Conventional alumina fiber is weaker than SiC or TiB_2, and it is heavier and a poorer thermal conductor than SiC. However, it stands out in other ways. Its heat resistance is exceptional, and fully oxidized Al_2O_3 will not degrade in oxygen at high temperatures. Its CTE closely matches that of titanium aluminides, and it does not react with gamma at high temperatures.

Al_2O_3 monofilament (Fiber FP) is produced with a slurry spinning process. Al_2O_3 particles and an aluminum-containing binder are mixed with a liquid to form a slurry with the consistency of cold maple syrup, are spun into fibers, and are charred or pyrolized in one continuous process. When heated, the aluminum in the binder becomes part of the fiber, organics vaporize, and the Al_2O_3 particles densify as they fuse together.

Al_2O_3 may also be a candidate to reinforce nickel aluminide, an intermetallic under consideration for turbines. Nickel aluminide does not ordinarily wet out alumina fibers, so the composites are not very tough. Intermetallic matrix wettability is improved by adding two strong oxidizers, titanium and yttrium (Ref 43).

Despite the lack of substantial ductility in most of the aluminides, efforts at using them as matrices in composites have continued through development and commercial production phases. A substantial improvement in high-temperature strength, combined with a higher stiffness and a lower density, can be achieved by reinforcing the aluminides with a ceramic fiber. Improvement in room-temperature toughness, in compounds that exhibit low room-temperature toughness, can also be accomplished by tailoring the fiber-matrix interface bond strength to promote fiber pull-out. Several techniques are now being employed to fabricate ceramic-fiber-reinforced aluminides: powder metallurgy, diffusion bonding, plasma spraying, and pressure casting.

Discontinuously reinforced IMCs have been produced by a variety of techniques, almost all of which rely on powder processing: rapid solidification processing (Ref 44), mechanical alloying (Ref 45), reactive sintering (Ref 46), reactive

HIP (Ref 46), cryomilling (Ref 47), and XD synthesis (Ref 42, 48). Rapid solidification processing and mechanical alloying typically produce dispersion strengthening with a relatively low volume fraction (<5%) of fine (<0.1 mm) second phase. Alternatively, ingot metallurgy can be used to produce XD composites containing up to 10 vol% of the reinforcement. For larger volume fractions, melt viscosity becomes a limitation.

IMC Matrices Processing

Powder metallurgy involves blending chopped fibers with either prealloyed powders, elemental powders, or a mixture of the two, followed by either hot pressing, HIP, or extrusion (Ref 49-60). Both hot pressing and extrusion have been employed by the researchers at Oak Ridge National Laboratory (Ref 49, 50) to consolidate Ni-24Al-0.24B and Ni-18.5Al-8Cr-0.4Zr-0.1B composites with chopped 20 mm diameter Fiber FP.

Diffusion Bonding. In the diffusion bonding technique, the matrix is first rolled into thin sheets. The ceramic fibers are then carefully placed in the desired orientation between the sheets. The sandwich, which consists of alternative layers of the reinforcement and the matrix, is compressed at high temperatures to force the solid matrix to flow around the fibers (Ref 61). However, IMCs are difficult to roll into thin sheets because of their brittleness, their ordered structure, and their low diffusion rates. Therefore, they can't easily be deformed to fill the gaps between the fibers. An alternative technique has been devised, known as the powder cloth process (Ref 62).

Powder Cloth Process. In this process, developed by NASA, the matrix phase is initially formed into a flexible, cloth-like sheet by combining the matrix powder with a fugitive organic binder and blending this mixture into a slurry with the aid of an organic wetting agent. As the wetting agent evaporates, the slurry is rolled into pliable sheets 0.2 mm thick. Powder cloth processing has been used to fabricate fiber-reinforced IMCs based on Ti_3Al (Ti-24Al-11Nb), NiAl, and FeAl (Fe-40Al) matrix materials (all compositions in atomic percent).

Foil/fiber processing is like the powder cloth process except that the matrix takes the form of a thin rolled sheet or foil. The fully dense material requires no binders. SiC-Ti_3Al composites have been produced using the process, and efforts are underway to thermomechanically process NiAl foils from consolidated melt-spun ribbon. Preliminary work has resulted in intermetallic foils 0.13 to 0.20 mm thick (Ref 63-67).

Thermal Spray. In arc and plasma spraying, a mandrel is wrapped with a precisely spaced array of fibers, then sprayed with a layer of matrix material. The sprayed matrix material mechanically bonds the fibers in place, forming a single-layer composite, or monotape (Ref 62, 68).

Pressure casting has been applied to IMCs recently (Ref 69). In this technique, the matrix alloy is placed inside a bottom-pour crucible positioned on top of a mold that contains the fiber preform. The matrix alloy and the fiber preform are heated simultaneously, but independently, to high temperatures. Once the desired fiber and matrix temperatures are reached, the molten metal is poured onto the fibers. Because molten metals do not often wet ceramic fibers, a positive pressure is required to force the molten metal to infiltrate the fiber preform. Infiltration of fiber preform in this technique has been achieved by the application of pressure by an inert gas (Ref 43, 70-74).

Self-propagating high-temperature synthesis (Ref 75-82) has been used as an alternative powder process applicable to Fe_3Al. This method uses an exothermic reaction between elemental powders to synthesize the intermetallic. Fully dense Fe_3Al, Fe_3Al containing chromium additions, and Al_2O_3-particulate-reinforced composites have been produced using SHS and consolidation by HIP or the Ceracon process.

Infiltration and squeeze casting uses titanium powders, crystalline Al_2O_3 fibers 300 to 500 mm long (Saffil fibers from Imperial Chemical Industries), and commercially pure aluminum. A preform (5 to 15 mm thick) is made by mixing 20 mm titanium powder particles with sufficient Al_2O_3 fibers to achieve a final 10 vol% for the reinforcement. The mixing is done in a solvent medium to form a suspension. Latex (starch) is added as a binder to the suspension. Pressing the suspension densifies the preform, and molten aluminum is then forced into the preheated porous preform by a squeeze casting technique. The final step is the all-important heat treatment, which reacts the aluminum with the titanium to form the $TiAl_3$ intermetallic. The heat treatment, conducted at 600 to 700 °C, also requires HIP to attain full reaction between the aluminum and titanium. Other stable fibers, such as Fiber FP or PRD-166, are being considered for future work to achieve optimum processing conditions. Additionally, the method could be modified to include continuous reinforcements and possible multicomponent systems.

One of the projected designs for the control surface assemblies of the national aerospace plane is to produce a welded or brazed honeycomb core structure of titanium aluminide foils. The structure is bonded on all sides to a thin skin of alpha to produce a strong, lightweight, heat-resistant structure. Alternative skin designs include a three-layer truss core of titanium aluminide and an actively cooled six-layer sandwich (Ref 83).

There are other potential uses for titanium aluminide foils. The bonding of several layers of foil wrapped around a mandrel may be a viable way to form high-strength, lightweight hollow shafts. These shafts could be used in engine systems for missiles or hypersonic vehicles. One of the current designs for wide-chord fan blades for high-thrust commercial aircraft engines uses activated diffusion bonding to seal a sandwich of titanium honeycomb and titanium panels (Ref 84).

New synthesis methods may be required for new composite materials such as FGMs (Ref 85-92), HYMATS (macrolaminates and metal-polymer hybrid materials, Ref 93), interpenetrating polymer networks (Ref 94), microinfiltrated macrolaminated composites (Ref 95), liquid crystal polymers (Ref 96), and IMCs (Ref 97). These may force development of previously uneconomical processes if they offer a technical solution for advanced systems capability. IMCs present opportunities for reducing the number of stages in turbine engines, and in so doing they may be economically beneficial even at a higher materials cost, because the smaller number of stages leads to greater economy of use. From nearly all aspects of research, IMCs offer great potential for structural applications; however, their widespread use remains a decade or more in the future.

Conclusion

Although the composites industry has lost the strong defense funding base that has driven its growth, the fundamental advantages of substituting composites for metals remain intact. Important environmental issues have spurred manufacturing toward reduced waste, longer component life, and weight-efficient systems.

With metals, we begin with well-defined materials having a well-developed database. Using these materials and traditional processes, we manufacture a structure. The process is linear and well understood. For the most part, we have been using this approach with composites, and this has resulted in higher costs to build databases, inefficient design, and less than optimal structures.

To make the best use of composites, we must rethink our approach to the design and manufacture of composite structures. Perhaps we should begin with the function and design the part to serve the function. Such designs would include concepts for manufacturing, especially joining, and selecting a material would be the *last* step (Ref 98-100).

REFERENCES

1. Composite Materials Tech. Roadmap, EM89-203, 22 May 1989, from Department of Defense Papers 339-2287 and 339-2376
2. W.H. Schwartz, Fasteners Hold Their Own, *Assembly Eng.*, April 1989, p 24, 25
3. N. Albee, Aspects of Mechanically Fastening Advanced Composites, *Adv. Compos.*, Nov-Dec 1988, p 54-60
4. Fasteners for Composites, *Aircr. Eng.*, July 1988, p 2, 3
5. L.H. McCarty, Lightweight Fasteners Reduce Unit Stresses on Composites, *Des. News*, 4 Sept 1989, p 186, 187
6. Manufacturers are Developing Improved Structural Systems, *Aviat. Week Space Technol.*, 12 Dec 1988, p 67-72

7. M.M. Schwartz, *Composite Materials Handbook*, 2nd ed., McGraw Hill, 1992
8. J. Rausch, Fastening Fundamentals for Fabricated Composites, *Assembly Eng.*, Oct 1988, p 34-36
9. N.K. Young, C.C. Chen, and W.L. Cheng, Joining Sections of Thick Fiberglass Laminate, *4th Ann. Conf. Advanced Composites* (Dearborn, MI), 13-15 Sept 1988, ASM International, Society of Plastics Engineers and Engineering Society of Detroit, p 209-216
10. L.K. English, Field Repair of Composite Structures, *Mech. Eng.*, Sept 1988, p 37-39
11. DuPont Enters European Composites Market, *Mach. Des.*, 12 Feb 1993, p 23
12. Automation in Aerospace Fastening, *Aerosp. Eng.*, Aug 1992, p 13-18
13. P.E. Teague, Fasteners that Mate with New Materials, *Des. News*, 21 Sept 1992, p 98-102
14. *Adhesive Bonding Handbook for Advanced Structural Materials*, ESA PSS 03 210 Issue 1, Feb 1990 Rev O., European Research & Technology Centre, Netherlands
15. Synthesis of Metallic Materials for Demanding Aerospace Applications Using Powder Metallurgy Techniques, *Proc. 1991 P/M in Aerospace and Defense Technologies Symp.* (Tampa, FL), 4-6 March 1991, Metal Powder Industries Federation, p 5-36
16. S. Komarneni, Nanocomposites, *J. Mater. Chem.*, Vol 2 (No. 12), 1992, p 1219-1230
17. R.S. Averback, "Nanophase Ceramics," ARO-25526, 10-MS , Illinois University at Urbana-Champaign, 21 May 1992
18. P.B. Messersmith and S.I. Stupp, Synthesis of Nanocomposites: Organoceramics, *J. Mater. Res.*, Vol 7 (No. 9), Sept 1992, p 2599-2611
19. T. Hirai and M. Sasaki, In situ Processing of Inorganic Composites by Chemical Vapor Deposition, *Advanced Structural Inorganic Composites*, P. Vincinzini, Ed., Elsevier Science, 1991, p 541
20. T. Hirai and T. Goto, *Bull. Jpn. Inst. Met.*, Vol 28, 1989, p 960
21. *Adv. Mater. Proc.*, Vol 143 (No. 3), March 1993, p 14
22. JPRS Rept.-JST-92-033, 22 Dec 1992
23. K.E. Drexler, *Nanosystems: Molecular Machinery, Manufacturing, and Computation*, John Wiley & Sons, 1992
24. K.E. Drexler, Molecular Engineering: An Approach to the Development of General Capabilities for Molecular Manipulation, *Proc. Nat. Acad. Sciences*, Vol 78 (No. 9), Sept 1981, p 5275-5278
25. W.B. Hillig, Ceramic Matrix Composites—Research and Development in Japan, Chapter 3, FR 3/91, ECS 8902528, *JTEC Panel Rept. on Advanced Composites in Japan*, p 61-85
26. N. Sata and J. Ikeuchi, *J. Ceram. Soc., Jpn., Int. Ed.*, 1987, p 95:208
27. A. Kawasaki and R. Watanabe, *In Sintering '87* (Tokyo), Vol 2, Elsevier, 1988
28. *Adv. Mater.*, Vol 14 (No. 8), 25 May 1992, p 2

29. M. Sasaki, Y. Wang, T. Hirano, et al., *J. Ceram. Soc. Jpn., Int. Ed.*, 1989, p 97:530
30. Y. Yamaoka, K. Ninomiya, K. Kosata, et al., Bonding of FGM to Actively Cooled Panel, *Tokyo FGM '91*, 8-9 Oct 1991, p 245-251, in Japanese; JPRS-92-025
31. T. Hirai and M. Sasaki, *Japanese Society of Mechnical Engineers Int. J.*, Vol 34, 1991, p 123
32. M. Sasaki and T. Hirai, Fabrication and Thermal Barrier Characteristics of CVD SiC/C Functionally Gradient Material, *Proc. 1st Int. Symp. on FGM* (Sendai), M. Yamanouchi, M. Koizumi, T. Hirai, et al., Ed., 1990, p 83
33. C. Fujiwara, S. Nagata, S. Kiyotoh, et al., Fabrication Process for Actively Cooled FGM Plate, *Tokyo '91*, 8-9 Oct 1991, p 235-243, in Japanese; JPRS-92-025
34. J.S. Moya, A.J. Sanchez-Herencia, J. Requena, et al., Functionally Gradient Ceramics by Sequential Slip Casting, *Mater. Lett.*, Vol 14 (No. 5, 6), July 1992, p 333-335
35. P.K. Brindley, P.A. Bartolotta, and S.J. Klima, "Investigation of a SiC/Ti-24Al-11 Nb Composite," TM-100956, National Aeronautics and Space Administration, 1988
36. J.R. Stephens, Intermetallic and Ceramic Matrix Composites for 815 to 1370°C Gas Turbine Engine Applications, *Int. Conf. Metal and Ceramic Matrix Composites: Processing, Modeling and Mechanical Behavior* (Anaheim, CA), 19-22 Feb 1990, R.B. Bhagat, A.H. Clauer, et al., Ed., TMS, p 3-12
37. L.J. Westfall, D.L. McDanels, and D.W. Petrasek, "Advanced Fiber Development for Intermetallic Matrix Composites," TM-102109, E 4878, National Aeronautics and Space Administration, June 1989
38. C.J. Smith and J.E. Johnson, Advanced Materials for 21st Century UBE, *34th Int.. Gas Turbine Conf.* (Toronto), 7 June 1989, American Society of Mechanical Engineers
39. H. Klein, "NASA High Temperature Advanced Propulsion Study," Interim Rept. presented at NASA Lewis Res. Ctr., Cleveland, OH, 25 April 1989
40. D.L. McDanels and J.R. Stephens, "High Temperature Engine Materials Technology—Intermetallics and Metal Matrix Composites," TM-100844, National Aeronautics and Space Administration, 1988
41. D.W. Petrasek, High Temperature Intermetallic Matrix Composites, *HITEMP Review—1988, Advanced High Temperature Engine Materials Technology Prog.*, CP-10025, National Aeronautics and Space Administration, Nov 1988, p 67-82
42. Y.W. Kim, Intermetallic Alloys Based on Gamma Titanium Aluminide, *J. Met.*, Vol 41 (No. 7), 1989, p 24-30
43. S. Nourbakhsh and H. Margolin, Fabrication of High Temperature Fiber Reinforced Intermetallic Matrix Composites, *Int. Conf. Metal and Ceramic Matrix Composites: Processing, Modeling and Mechanical Behavior* (Anaheim, CA), 19-22 Feb 1990, R.B. Bhagat, A.H. Clauer, et al., Ed., TMS, p 75-90
44. J.D. Whittenberger, D.J. Gaydosh, and K.S. Kumar, *J. Mater. Sci.*, Vol 25, 1990, p 2771

45. J.D. Whittenberger, *Solid State Powder Processing*, A.H. Clauer and J.J. deBarbadillo, Ed., The Minerals, Metals and Materials Soc., 1990, p 137
46. N.S. Stoloff and D.E. Alman, *Intermetallic Matrix Composites*, D.L. Anton et al., Ed., Vol 194, Materials Research Society, 1990, p 31
47. J.D. Whittenberger, E. Arzt, and M.J. Luton, *Intermetallic Matrix Composites*, D.L. Anton et al., Ed., Vol 194, Materials Research Society, 1990, p 221
48. J.M. Brupbacher, L. Christodoulou, and D.C. Nagle, U.S. Patent No. 4,710,348, 1 Dec 1987
49. G.L. Povirk, J.A. Horton, C.G. McKamey, et al., Interfaces in Nickel Aluminide/Alumina Fibre Composites, *J. Mater. Sci.*, Vol 23, 1988, p 3945-3950
50. C.G. McKamey, G.L. Povirk, J.A. Horton, et al., Fabrication and Mechanical Properties of Ni_3Al-Al_2O_3 Composites, *High-Temperature Ordered Intermetallic Alloys III*, C.T. Liu, A.I. Taub, N.S. Stoloff, et al., Ed., Vol 133, Materials Research Society, 1989, p 609-614
51. A. Bose and R.M. German, Novel Processing Approaches to Intermetallic Matrix Composites, *Adv. Mater. Proc.*, Vol 3, 1988, p 37-58
52. D.E. Alman and N.S. Stoloff, Fabrication and Mechanical Properties of Powder Injection Molded Intermetallic Matrix Composites, *Composite Materials, Mechanics and Processing*, Proc. 6th Technical Conf., American Society for Composites, Technomic Publishing, p 390-399
53. D.E. Alman and N.S. Stoloff, Powder Fabrication of Monolithic and Composite NiAl, *Int. J. Powder Metall.*, Vol 27 (No. 1), 1991, p 29-41
54. N.S. Stoloff and D.E. Alman, Powder Processing of Intermetallic Alloys and Intermetallic Matrix Composites, *Mater. Sci. Eng. A*, Vol 144, 1991, p 51-62
55. S.E. Mukherjee and G.P. Khandra, Development of Nickel Aluminide-Carbon Fibre Intermetallic Composite by Reaction Sintering, *J. Mater. Sci. Lett.*, Vol 10, 1991, p 1222-1224
56. D.E. Alman, R.M. Cooper, and K.G. Shaw, Nickel Aluminide Intermetallics as a Matrix for Diamonds in Cutting Tools, *Composite Materials, Mechanics and Processing*, Proc. 6th Technical Conf., American Society for Composites, Technomic Publishing, p 648-658
57. D.J. Larsen and L.V. Interrante, Application of Chemical Vapor Deposited Yttria for the Protection of Silicon Carbide Fibers in a SiC/Ni_3Al Composite, *J. Mater. Res.*, Vol 5 (No. 11), Nov 1990, p 2706-2717
58. W.E. Misiolek and R.M. German, Fabrication of NiAl/TiB_2 Intermetallic Matrix Composites, *Proc. of 1990 P/M Conf. and Exhibition* (Pittsburgh, PA), 20-23 May 1990, E.R. Andreotti and P.J. McGeehan, Ed., Vol 2, Metal Powder Industries Federation and American Powder Metallurgy Institute, p 161-172
59. S.M. Barinov, V.Yu. Evdokimov, and V.Yu. Shevchenko, Zirconia-Toughened Nickel Aluminide, *J. Mater. Sci. Lett.*, Vol 11, 1992, p 1347-1348
60. H. Chen, M. Kaya, and R.W. Smith, Near Net-Shape Long Fibre Reinforced Intermetallic Matrix Composites Produced by Reactive Infiltration Process, *Mater. Lett.*, Vol 13 (No. 4, 5), April 1992, p 180-183

61. T.W. Chou, A. Kelly, and O. Okura, Fibre-Reinforced Metal-Matrix Composites, *Composites*, Vol 16 (No. 3), 1985, p 187-206
62. R. Bowman and R. Noebe, Up-and-Coming IMCs, *Adv. Mater. Proc.*, Vol 136 (No. 2), 1989, p 35-40
63. S.C. Jha, J.A. Foster, A.K. Pandey, et al., MMCs via Titanium Aluminide Foils, *Adv. Mater. Proc.*, Vol 139 (No. 4), p 87-90
64. M. Khobaib, Damage Evolution in Creep of SCS 6/Ti-24Al-11Nb Metal Matrix Composites, *Composite Materials, Mechanics and Processing*, Proc. 6th Technical Conf., American Society for Composites, Technomic Publishing, p 638-647
65. S.M. Jeng, J.-M. Yang, D.G. Rosenthal, et al., Mechanical Behavior of SiC Fibre-Reinforced Titanium/Titanium Aluminide Hybrid Composites, *J. Mater. Sci.*, Vol 27, 1992, p 5357-5364
66. "Titanium Aluminide/Titanium Composite," U.S. Patent No. 5,102,451
67. D.L. McDanels and L.J. Westfall, High Temperature Intermetallic Matrix Composites—Fiber Development, *HITEMP Review—1988, Advanced High Temp. Engine Materials Technology Prog.*, CP-10025, National Aeronautics and Space Administration, Nov 1988, p 205-214
68. R. Tiwari, H. Herman, S. Sampath, et al., Plasma Spray Consolidation of High Temperature Composites, *Mater. Sci. Eng. A*, Vol 144, 1991, p 127-131
69. S. Nourbakhsh, F.L. Liang, and H. Margolin, An Apparatus for Pressure Casting of Fibre-Reinforced High-Temperature Metal-Matrix Composites, *J. Phys. E.; Sci. Instrum.*, Vol 21, 1988, p 898-902
70. E.A. Feest and J.H. Tweed, Comparative Viability of Processing Routes for Intermetallic Based Materials, *Mater. Sci. Technol.*, Vol 8, April 1992, p 308-316
71. S. Nourbakhsh, O. Sahin, W.H. Rhee, et al., Microstructural Characterization of a Zirconia-Toughened Alumina Fiber Reinforced Niobium Aluminide Composite, *Acta Metall. Mater.*, Vol 40 (No. 2), 1992, p 285-294
72. S. Nourbakhsh, H. Margolin, and F.L. Liang, Microstructure of a Pressure-Cast Fe_3Al Intermetallic Alloy Composite Reinforced with Zirconia-Toughened Alumina Fibers, *Metall. Trans. A*, Vol 21, Nov 1990, p 2881-2889
73. S. Nourbakhsh and H. Margolin, Processing of Continuous-Ceramic-Fiber-Reinforced Intermetallic Composites by Pressure Casting, *Mater. Sci. Eng. A*, Vol 144, 1991, p 133-141
74. S. Nourbakhsh, W.H. Rhee, O. Sahin, et al., Mechanical Behavior of a Reinforced Ni_3Al Matrix Composite, *Mater. Sci. Eng. A*, Vol 153, 1992, p 619-627
75. B.H. Rabin, R.N. Wright, J.R. Knibloe, et al., Reaction Processing of Iron Aluminides, *Mater. Sci. Eng. A*, Vol 153, 1992, p 706-711
76. SHS Forms Layered Composites, *Adv. Mater. Proc.*, Vol 142 (No. 4), Oct 1992, p 9
77. D.M. Sims, A. Bose, and R.M. German, *Proc. Powder Metall.*, Vol 43, 1987, p 575
78. A. Bose, B.H. Rabin, and R.H. German, *Powder Metall. Int.*, Vol 20, 1988, p 25

79. C.G. McKamey, J.A. Horton, and C.T. Liu, *J. Mater. Res.*, Vol 4, 1989, p 1156
80. J.R. Knibloe, R.N. Wright, and V.K. Sikka, *Advances in Powder Metallurgy*, Vol 2, Metal Powder Industries Federation, 1990, p 219
81. Z. Eliezer, B.-H. Lee, C.J. Hou, et al., Matrix-Reinforcement Interface Characteristics of (Ti_3Al + Nb)-Based Powder Composites, Consolidated by High-Energy High-Rate Processing, *Int. Conf. Metal and Ceramic Matrix Composites: Processing, Modeling and Mechanical Behavior* (Anaheim, CA), 19-22 Feb 1990, R.B. Bhagat, A.H. Clauer, et al., Ed., TMS, p 401-412
82. J.C. Rawers, W.R. Wrzesinski, E.K. Roub, et al., TiAl-SiC Composites Prepared by High Temperature Synthesis, *Mater. Sci. Technol.*, Vol 6, Feb 1990, p 187-191
83. R.N. Boggs, *Des. News*, Vol 6 (No. 19), 1989, p 51-53
84. G. Norris, *Flight Int.*, May 1990, p 34-38
85. W.B. Hillig, Ceramic Matrix Composites—Res. and Dev. in Japan, Chapter 6.3, JTEC Panel Rept. on Advanced Composites in Japan, FR 3/91, ECS 8902528, p 61-85
86. N. Sata and J. Ikeuchi, *J. Ceram. Soc., Jpn., Int. Ed.*, 1987, p 95:208
87. A. Kawasaki and R. Watanabe, *In Sintering '87* (Tokyo), Vol 2, Elsevier, 1988
88. M. Sasaki, Y. Wang, T. Hirano, et al., *J. Ceram. Soc., Jpn., Int. Ed.*, 1989, p 97:530
89. T. Hirai and M. Sasaki, *JSME Int. J.*, Vol 34, 1991, p 123
90. M. Sasaki and T. Hirai, Fabrication and Thermal Barrier Characteristics of CVD SiC/C Functionally Gradient Material, *Proc. 1st Int. Symp. FGM* (Sendai), M. Yamanouchi, M. Koizumi, T. Hirai, et al., Ed., 1990, p 83
91. "New Trends for Advanced Materials," Selections from Proceedings of Symposium on Science and Technology: Japan, Joint Publications Research Service, JPRS-JST-92-012, 10 April 1992
92. J.S. Moya, A.J. Sanchez-Herencia, J. Requena, et al., Functionally Gradient Ceramics by Sequential Slip Casting, *Mater. Lett.*, Vol 14 (No. 5, 6), July 1992, p 333-335
93. W.E. Frazier, M.E. Donnellan, P. Architetto, et al., The Status of HYMATs—A New Category of Hybrid Materials, *JOM*, Vol 43 (No. 5), May 1991, p 10-15
94. D.R. Clarke, Interpenetrating Phase Composites, *J. Am. Ceram. Soc.*, Vol 75 (No. 4), 1992, p 739-759
95. A. Bose and J. Lankford, MIMLCs: New Composite Architecture, *Adv. Mater. Proc.*, Vol 140 (No. 1), July 1991, p 18-22
96. M. Hunt, Star Wars Materials Launch Commercial Products, *Mach. Des.*, 26 March 1993, p 52-58
97. JPRS Symposium on High-Performance Materials: Sci. & Technology: Japan Basic Technologies for Future Industries, Selected papers from 2nd Mtg., JST-93-006-L, 25 Jan 1993
98. J.K. Lees, Composites Persist despite Funding Cuts, *Am. Ceram. Soc. Bull.*, Vol 72 (No. 8), Aug 1991, p 99

99. J.K. Lees and S.J. Wilcox, Make the Most of Composites, *Des. News*, 18 May 1992
100. F. Kodama, Technology Fusion and the New, *Harvard Bus. Rev.*, July / Aug 1992

SELECTED REFERENCES

- 109th Conf., JPRS-JST-92-058-L, Japan Institute of Metals, Sci. & Tech., 23 July 1992
- Advanced Composite Materials 1992 Seminar; Sci. & Technology: Japan, JPRS-JST-93-007, March 1993
- S. Ashley, Titanium Aluminides: Tough Materials for Tomorrow's Engines, *Mech. Eng.*, Vol 113 (No. 12), Dec 1991, p 48-52
- A.A. Baker, R.J. Callinan, M.J. Davis, et al., Repair of Mirage III Aircraft Using the BFRP Crack-Patching Technique, *Theor. Appl. Frac. Mech.*, Vol 2, 1984, p 1-1S
- J.D. Buckley, 15th Conf. on Metal Matrix, Carbon, and Ceramic Matrix Composites (Cocoa Beach, FL), 16-18 Jan 1991, CP-3133, Part 1, National Aeronautics and Space Administration
- W.A.T. Clark and J.A. Moser, "Compatibility of Potential Reinforcing Ceramics with Ni and Fe Aluminides," NAS 1:26:192232, CR-192232, National Aeronautics and Space Administration, 10 Dec 1991
- E.L. Courtright, H.C. Graham, A.P. Katz, et al., "Ultra-High Temperature Assessment Study—Ceramic Matrix Composites," U.S. Government Report AD-A262 740, WL-TR-914061, Sept 1992
- Dental Implant Using FGM, *New Technol. Jpn.*, Vol 20 (No. 11), Feb 1993, p 17
- D.M. Dimiduk, D.B. Miracle, and C.H. Ward, Development of Intermetallic Materials for Aerospace Systems, *Mater. Sci. Technol.*, Vol 8 (No. 4), April 1992, p 367-375
- H. Eschbaumer, "Repair Procedures for Advanced Composites for Helicopters," ERF9 1-88, 23p, MBB-UD-0606-91-PUB 17, European Rotorcraft Forum (Berlin, Germany), 24-27 Sept 1991
- W.F. Gale and S.V. Orel, "Diffusion Brazing of Nickel Aluminides," Session B3, Paper B3C, 25th AWS Int. Brazing and Soldering Conf., 13 April 1994 (Philadelphia, PA)
- D.M. Ginburg, Metal Matrix Composite Vertical Tail Fabrication, EM90-438, *SME Metal Matrix Clinic* (Anaheim, CA), 13-14 Nov 1990
- J. Kesapradist, K. Ono, and K. Fukaura, Producing Ni_3Al Matrix Composite Material by Vacuum Hot Pressing and Heat Treatment of Nickel-Plated Aluminum Sheets with Al_2O_3 Fiber, *Mater. Sci. Eng. A*, Vol 153 (No. 1-2), 1992, p 641-645

- J.-P. Lebrat, A. Varma, and A.E. Miller, Combustion Synthesis of Ni_3Al and Ni_3Al and Ni_3Al-Matrix Composites, *Metall. Trans. A*, Vol 23 (No. 1), Jan 1992, p 69-76
- E.U. Lee, T. Kirscher, and J. Waldman, "Oxidation of Titanium Aluminide and Its XD Composite," Final Rept., NADC-90122-60, Naval Air Development Center, Warminster, PA, Dec 1990
- S.M. Lee, Part II—Nanocomposites A Technology Driven to Realization, *SAMPE J.*, Vol 29 (No. 3), May/June 1993, p 2
- X. Liang, H.K. Kim, J.C. Earthman, et al., Microstructure and Elevated Temperature Behavior of Spray-Atomized and Co-Deposited $Ni_3Al/SiC/TiB_2$ Intermetallic Matrix Composites, *Mater. Sci. Eng. A*, Vol 153 (No. 1-2), 1992, p 646-653
- T.J. Lienert, W.A. Baeslack, III, M. Kuo, et al., "Inertia Friction Welding of SiC-Reinforced 8009 Aluminum," Session 16, Paper A, 75th AWS Int. Ann. Mtg. Conf. (Philadelphia), 14 April 1994
- J.Y. Liu and B.A. Chin, Session B2, Paper B2A, 25th AWS Int. Brazing and Soldering Conf. (Philadelphia), 13 April 1994
- J.E. McCarthy, *Composites*, Vol 24 (No. 4), 1993, p 361-365
- A.K. Miller and G.S. Springer, "Enabling Technologies for Economical Manufacturing of Composites," AD-A253 615, July 1992
- S. Musicant, *Space Shuttle Insulation Tiles*, Marcel Dekker, p 147-151
- S. Naka, M. Thomas, and T. Khan, Potential and Prospects of Some Intermetallic Compounds for Structural Applications, *Mater. Sci. Technol.*, Vol 8 (No. 4), 1992, p 291-298
- V.R. Parmeswaran, High-Temperature Aluminides and Intermetallics, *J. Adhes.*, Vol 8 (No. 1-2), 1992, p 41-43
- R. Raj, "Nanograin Superplasticity in Ceramics," AD-A256 643, Sept 1992
- F.A. Sandow and R.K. Cannon, "Composite Repair of Cracked Aluminum Alloy Aircraft Structure," Report AFWAL-TR-87-3072, Air Force Wright Aeronautical Laboratories, 1987
- L.M. Sheppard, Enhancing Performance of Ceramic Composites, *Am. Ceram. Soc. Bull.*, Vol 71 (No. 4), 1992, p 617-632
- V.K. Sikka, J.T. Mavity, and K. Anderson, Processing of Nickel Aluminides and Their Industrial Applications, *Mater. Sci. Eng. A*, Vol 153, 1992, p 712-721
- M.J. Strum and G.A. Henshall, "Diffusion Bonding of NiAl with Ni and Al Interlayers," Session 3, Paper C, 75th AWS Int. Ann. Mtg. Conf. (Philadelphia), 11 April 1994
- M. Takahashi, Y. Itoh, and H. Kashiwaya, New Ways to Make "Functionally Gradient Materials," *Weld. J.*, March 1992, p 93
- K. Tangri, "Development of Ductilized Titanium Aluminides as a Matrix for High Temperature, High Performance Composites," DREP-91-20, CTN-92-60581, Manitoba University, June 1991

- TiAlNb-Based Alloys Outperform Conventional Titanium Aluminides, *Adv. Mater. Proc.*, Vol 141 (No. 3), March 1992, p 33-35
- P.A. Walls and M. Ueki, Joining SiALON Ceramics Using Composite β-SiALON-Glass Adhesives, *J. Am. Ceram. Soc.*, Vol 75 (No. 9), 1992, p 2491-2497
- S.H. Whang, C.T. Liu, D.P. Pope, et al., High Temperature Aluminides and Intermetallics, *Proc. of 1989 Symposium on: TMS Fall Mtg.*(Indianapolis), 1-5 Oct 1989
- M. Yamaguchi, High Temperature Intermetallics—with Particular Emphasis on TiAl, *Mater. Sci. Technol.*, Vol 8, April 1992, p 299-307

Index